Radomir S. Stanković and Jaakko Astola

From Boolean Logic to Switching Circuits and Automata

Studies in Computational Intelligence, Volume 335

Editor-in-Chief
Prof. Janusz Kacprzyk
Systems Research Institute
Polish Academy of Sciences
ul. Newelska 6
01-447 Warsaw
Poland
E-mail: kacprzyk@ibspan.waw.pl

Further volumes of this series can be found on our homepage: springer.com

Vol. 313. Imre J. Rudas, János Fodor, and Janusz Kacprzyk (Eds.)
Computational Intelligence in Engineering, 2010
ISBN 978-3-642-15219-1

Vol. 314. Lorenzo Magnani, Walter Carnielli, and Claudio Pizzi (Eds.)
Model-Based Reasoning in Science and Technology, 2010
ISBN 978-3-642-15222-1

Vol. 315. Mohammad Essaaidi, Michele Malgeri, and Costin Badica (Eds.)
Intelligent Distributed Computing IV, 2010
ISBN 978-3-642-15210-8

Vol. 316. Philipp Wolfrum
Information Routing, Correspondence Finding, and Object Recognition in the Brain, 2010
ISBN 978-3-642-15253-5

Vol. 317. Roger Lee (Ed.)
Computer and Information Science 2010
ISBN 978-3-642-15404-1

Vol. 318. Oscar Castillo, Janusz Kacprzyk, and Witold Pedrycz (Eds.)
Soft Computing for Intelligent Control and Mobile Robotics, 2010
ISBN 978-3-642-15533-8

Vol. 319. Takayuki Ito, Minjie Zhang, Valentin Robu, Shaheen Fatima, Tokuro Matsuo, and Hirofumi Yamaki (Eds.)
Innovations in Agent-Based Complex Automated Negotiations, 2010
ISBN 978-3-642-15611-3

Vol. 320. xxx

Vol. 321. Dimitri Plemenos and Georgios Miaoulis (Eds.)
Intelligent Computer Graphics 2010
ISBN 978-3-642-15689-2

Vol. 322. Bruno Baruque and Emilio Corchado (Eds.)
Fusion Methods for Unsupervised Learning Ensembles, 2010
ISBN 978-3-642-16204-6

Vol. 323. Yingxu Wang, Du Zhang, and Witold Kinsner (Eds.)
Advances in Cognitive Informatics, 2010
ISBN 978-3-642-16082-0

Vol. 324. Alessandro Soro, Vargiu Eloisa, Giuliano Armano, and Gavino Paddeu (Eds.)
Information Retrieval and Mining in Distributed Environments, 2010
ISBN 978-3-642-16088-2

Vol. 325. Quan Bai and Naoki Fukuta (Eds.)
Advances in Practical Multi-Agent Systems, 2010
ISBN 978-3-642-16097-4

Vol. 326. Sheryl Brahnam and Lakhmi C. Jain (Eds.)
Advanced Computational Intelligence Paradigms in Healthcare 5, 2010
ISBN 978-3-642-16094-3

Vol. 327. Slawomir Wiak and Ewa Napieralska-Juszczak (Eds.)
Computational Methods for the Innovative Design of Electrical Devices, 2010
ISBN 978-3-642-16224-4

Vol. 328. Raoul Huys and Viktor K. Jirsa (Eds.)
Nonlinear Dynamics in Human Behavior, 2010
ISBN 978-3-642-16261-9

Vol. 329. Santi Caballé, Fatos Xhafa, and Ajith Abraham (Eds.)
Intelligent Networking, Collaborative Systems and Applications, 2010
ISBN 978-3-642-16792-8

Vol. 330. Steffen Rendle
Context-Aware Ranking with Factorization Models, 2010
ISBN 978-3-642-16897-0

Vol. 331. Athena Vakali and Lakhmi C. Jain (Eds.)
New Directions in Web Data Management 1, 2011
ISBN 978-3-642-17550-3

Vol. 332. Jianguo Zhang, Ling Shao, Lei Zhang, and Graeme A. Jones (Eds.)
Intelligent Video Event Analysis and Understanding, 2011
ISBN 978-3-642-17553-4

Vol. 333. Fedja Hadzic, Henry Tan, and Tharam S. Dillon
Mining of Data with Complex Structures, 2011
ISBN 978-3-642-17556-5

Vol. 334. Álvaro Herrero and Emilio Corchado
Mobile Hybrid Intrusion Detection, 2011
ISBN 978-3-642-18298-3

Vol. 335. Radomir S. Stanković and Jaakko Astola
From Boolean Logic to Switching Circuits and Automata, 2011
ISBN 978-3-642-11681-0

Radomir S. Stanković and Jaakko Astola

From Boolean Logic to Switching Circuits and Automata

Towards Modern Information Technology

Springer

Prof. Dr. Radomir S. Stanković
Dept. of Computer Science
Faculty of Electronics
University of Niš
Aleksandra Medvedeva 14
18000 Niš
Serbia
E-mail: rstankovic@bankerinter.net
Radomir.Stankovic@gmail.com

Prof. Dr. Jaakko Astola
Dept. of Signal Processing
Tampere University of Technology
P.O. Box 553
FI-33101Tampere
Finland
E-mail: Jaakko.Astola@tut.fi

ISBN 978-3-642-11681-0 e-ISBN 978-3-642-11682-7

DOI 10.1007/978-3-642-11682-7

Studies in Computational Intelligence ISSN 1860-949X

Library of Congress Control Number: 2011921126

Typeset & Cover Design: Scientific Publishing Services Pvt. Ltd., Chennai, India.

Printed on acid-free paper

9 8 7 6 5 4 3 2 1

springer.com

Contents

Part I: Mathematical Foundations

1 From Logic to Mathematical Logic ... 3
- 1.1 Revival of Study of Logic in Britain ... 4
- 1.2 Elements of Logic by Whately ... 6
- 1.3 Ten Rules of Philosophizing by Herschel ... 14
- 1.4 Inductive Science by Whewell ... 18
- 1.5 Mill and a System of Logic ... 25

2 Boolean Logic ... 35
- 2.1 Formal Logic by De Morgan ... 38
- 2.2 Work on Logic by Boole ... 46
- 2.3 Mathematical Analysis of Logic ... 47
- 2.4 The Calculus of Logic ... 51
- 2.5 The Laws of Thought ... 53
- 2.6 Boolean Algebra ... 64
- 2.7 Work on Logic by Jevons ... 66
- 2.8 Jevons and Boole ... 71
- 2.9 Mechanization of Boolean Logic ... 72
- 2.10 Contributions of P.S. Poreckij ... 84

3 Algebra of Logic ... 87
- 3.1 Algebra of Logic by Schröder ... 91
- 3.2 Contributions by Ch. S. Peirce ... 106
- 3.3 Venn Diagrams ... 110

4 Boolean Algebra – A Mathematical Subject ... 111
- 4.1 Axiomatization by Huntington ... 114
- 4.2 Boolean Algebra – A Strong Mathematical Subject ... 117

Part II: Switching Theory

5 Application of Algebra of Logic in Engineering 121
5.1 Remarks and Ideas by P. Ehrnfest 124

6 Switching Theory – From Art and Skills to Scientific Methods 129
6.1 Switching Theory from Boolean Algebra 132
6.2 Review of the Work by Shannon 137
6.3 Review by Baylis 139

7 Switching Theory in USSR 141
7.1 Pioneering Work by Šhestakov 144
7.2 Contributions by Other Researchers 149

8 Switching Theory in Japan 153
8.1 Akira Nakashima 156
8.2 Nakashima and Hanzawa 163
8.3 Continuation of Research 166

9 Switching Theory in Europe 167
9.1 Johanna Piesch 170
9.2 Vienna School in Switching 174

10 Postscript 183
10.1 Switching Theory 183
10.1.1 Boolean Algebra 184
10.2 Algebra of Logic 185
10.2.1 Work by G.J. Boole 185
10.2.2 Work by Platon S. Poreckij 187
10.3 Applications of Boolean Algebra 188
10.3.1 First Applications of Algebra of Logic in Circuit Design 188
10.3.2 Work by V.I. Šestakov 188
10.3.3 Work by C.E. Shannon 189
10.3.4 Work by A. Nakashima 191
10.3.5 Work by J. Piesch 193

References in the Postscript 195

List of Illustrations 207

Preface

This book contains travelers logs and records of milestones on the road we have traveled to understand the development of *Switching Theory*.

Besides being a reminder of an exciting journey, we believe it can be useful for other travelers to the same spiritual territory or perhaps some other destinations by making different choices at various cross-roads. If so, we would be glad, and will have an excuse to presenting it to public notice.

Voyagers,

Niš, Serbia, Tampere, Finland, *Radomir Stanković, Jaakko Astola*

August 2010

How to Use This Book

Dear Traveler Companion,

This journey to the *History of Switching Theory* has few sections. A traveler may choose one of them and go along by exploring presented details and contemplating them. Then, he may decide to take another part and continue. Please, pay attention to selected pages and read text on them as one reads inscriptions on famous monuments. These pages are really milestones and monuments in the imaginary Word that we are traveling through. Look into distribution of other details and their placement, some discrete colours and lines, as on traffic lights and road signals. Then, when the journey is completed, you may want to read the Postscript to compare you impressions and conclusions with those we got along the same trip.

Someone who carefully plans ones journeys may want to first read the Postscript as a guide book, carefully select places to visit, and then invest time and efforts in exploring them.

Whatever is the way you would chose, we wish you a nice trip.

Срећан пут
Hyvä matka

Intentions

Studying the historical development of theoretical foundations and tracing the train of thought that led to their formulations and to related practical discoveries are very suitable ways of learning and getting insight into how to determine the roads for future research and investigations in similar areas.

This booklet presents some historical remarks about the development of the switching theory and recalls and highlights some not widely known details about the contributions to this area of scholars from different parts of the world. In particular, we discuss the origins of the application of Boolean algebra in circuit design and related topics.

This booklet contains many reprints of pages selected from important publications by several authors. We did not want to rephrase or rewrite their original statements, since we believe that the way they were presented originally by authors, has a particular value for the reader. The manner of writing and the way of presenting some important concepts often reflects not just the fashion of the times, but also the authors attitude to the subject being discussed and to the other circumstances at the time when the notions were introduced and theories formulated.

Therefore, we did not dare disturb and destroy the initial composition of the contents and their formulation, as done by the authors, by rewording the statements in present terminology. Thus, we restricted the contribution on our part to the selection of what we consider most interesting to present to the readers notice. We kindly ask for these reprints not to be considered simply as graphic illustrations from previous publications, but to be read as a part of the presentations.

We have also kept the hand written remarks which we wrote for ourselves when recording a particular page for selection, since they explain in brief the motivation for each page selection.

About the Way of Writing

Advanced technologies for transmitting and storing information drastically change also the way of presenting it to potential users. Especially the Internet, as a relatively new medium for presenting and reaching information has a very strong impact to the way of anticipating facts and getting knowledge, as well as the manner of learning in general.

Very often, combinations of pictures, diagrams and drawings with text, expressed in short segments that the screen space allows, make this way of transferring knowledge much more appealing for customers, formerly readers, than whatever smart and deep considerations when presented in the form of a condense typed text in classical book formats.

Dynamic way of leaving and professional work usually shortly narrow towards specialized areas, often do not allow time for learning aside topics although as individuals we may be highly interesting in that.

In such circumstances, due to the lack of time for a longer continuous study of a particular subject, many readers prefer to get some facts in an obvious as possible manner, simply to anticipate, quick to learn, and easy to remember. This way of instant learning is quite accepted, and allows an individual to increase knowledge although the information is presented fragmentary, but allows him to fill the gaps by himself, learning piecewise, whenever possible. Sometimes, we find the linking information unexpectedly when looking for various, quite unrelated information by using advanced technology, just due to the imperfection of it. It means, interesting links and web pages can appear due to imperfection of tagging and similar reasons although we were not looking for them at the particular moment. Then, the way of thinking may go in different directions than previously planned resulting in widening the horizons.

Short web articles written and posted by many enthusiastic individuals, as well as various free encyclopedia, get their popularity, among other reasons, due to fitting well into this trend of learning and self-education noticeable nowadays.

As eager users of new technologies, but still funs of classical books, and all their strong features, we believe a compromise can be made by slightly changing the way of writing books to make them more appealing for potential readers, when the subjects considered allows to do that.

We believe that, when possible, books should be short. It is advisable that the presentation is split into reasonably small but self-contained bibliographic items that might be later assembled into larger and more complete volumes.

This may bring a potential reader easier to the conclusion as, Well I may find time to go it through, instead just putting the book aside after some possibly highly interesting look over it, waiting for some more free time, that hardly ever comes.

Furthermore, the presentation should be more likely as that the users of other media are customized. This is especially important for the youngster generation of potential readers, that first start with electronic media and spend much more time in joining with them than that devoted to classical books.

The presentation should be open in the sense that the reader should feel welcomed to participate into further development of the book by introducing by himself missing parts between carefully selected basic elements as fundamental building blocks of the presented knowledge. It is assumed that some necessary remarks and guide lines, that we call mile stones in the Preface, are provided for a proper navigation of the reader. These remarks should be short and highly informative, and in this respect can be compared to the traffic signs along a road.

However, selected references for further reading should be provided as a guidance to those that may want to explore the subject further and invest more time in study of it.

The selection of facts to be presented, and the way of linking them or putting together each other, should provide a vehicle to escape of being superficial in a study and to avoid presenting just a grasp of facts collected without deep thinking of relationships among them.

This booklet is an attempt in this direction and we are aware that it can be criticized in many aspects as well as the general idea behind witting, better to say drawing, or simply preparing it. Any criticism towards improving it, as well as concrete suggestions how to do that, are very welcome.

Differences in the Way of Writing

We will use the page 4 of the present book as an example illustrating what the authors of this book mean under the new way of writing, which also necessarily implies the new way of reading books, and that are both motivated by the advent of new electronic media.

Classical Way of Writing

In the classical way of writing the contents of the page can be presented as follows.

Although methods of logic and were obviously present in many cultures, which all used some intricate systems of reasoning, it is commonly accepted that explicit analysis of the principles of reasoning were developed independently in China, India, and Greece. The later being the most influential to the systems of logic in the West. In particular, the Aristotelian logic was widely accepted in the western science and mathematics. Many scholars contributed to the development, which has been continued by Islamic and medieval European scholars. The mid-fourteenth century is considered to be the period of the most respectable achievements. The period that followed including the first three or four decades of nineteenth century is viewed as a barren or even period of declination and degradation of logic.

There has been apparent a revival of study in Logic in Great Britanin at the beginning of 19th century. The most recognized scholars at that period of the time in this area were Richard Whately, John F.W. Herschel, William Whewell, and John Stuart Mill. They published considerable contributions in 1826, 1830, 1837 and 1840, and 1843, respectively. Their work leaded to the establishment of the scientific area that is nowadays called the *intuitive logic*. This revival of the interest in study of logic set a good background and provided a suitable milieu for the later work by Augustus De Morgan and George J. Boole, whose independent publications in 1847 set foundations of the scientific area called the *symbolic logic*.

Suggested Way of Writing

We first put the title and it clearly suggests that there has been a revival in study in logic in Great Britain to focus the attention of the reader to this fact.

Then, we put the block briefly describing the situation in logic before that, since it is supposed that this is already known by the reader.

Then, we write names four important scholars and posted them at the page together within a small spatial distance, and we add years to each of them.

The goal is twofold. Writing the names together immediately suggests that their work is related. These years associated indicate the period of the revival of the interest, and the reader is able to make a right conclusion about that. At the same time, the reader will wonder what these years more concretely mean, why exactly these years are pointed out, and will continue searching an answer to that. This motivates him for further reading.

Then, we put a vertical line to connect the names, and added the words intuitive logic. It is supposed that the reader will easily conclude that this is the result of the work of the scholars connected by the line and it is called the intuitive logic.

The names of two other scholars are grouped separately and separated form the previous group by a white space. This indicates that they do not belong to the same group, but since the their names are rather closely located at the same page, the conclusion should be that there are some links between. Since, the same year is repeated twice, the conclusion is that the work by these scholars has been given independently, otherwise the grouping of names will be different, in the same row, and the year will not be repeated. The explanations of these guessing are given at pages that follow. This should be more intriguing for further reading than simple writing down all of that.

Another example of the manner of writing adopted for this book is the following. We selected a page in a very important publication by George J. Boole, where he for the first time had used the equation $x \cdot x = x$, that is a basic building block in the beautiful construction of Boolean logic. We reprinted this page (at the page 62) and put a remark for the reader explaining why this particular page is selected. This remark has the role of a traffic sign pointing to an important turnover on the road towards the mathematical logic. We kindly suppose that the reader will stop and read the entire text on the selected page and get both the information on the subject as well as some insight into the way of presenting scientific work at the time, which has it own value.

We do the similar with pages where references to the previous and related work were given by the respective authors. Pointing these references, we offer to the reader a necessary information about the links to the background work as viewed by the authors of the corresponding publications, as well as how they estimated relationships of their own work to that of predecessors or contemporarily active scholars.

This manner of presenting important publications in the area will be used in this book whenever appropriate.

Why a New Way of Writing

There can be expressed several rationales why the new way of writing, if it is the new at all or just a bit different, might be worth of suggesting. Few of them we may want to point are the following.

1. *To give the reader a more active role.*

Instead of being a passive subject of tutoring, exposed to all the conclusions explicitly presented in the way as the author formulated them, the reader is put into position to draw conclusion by himself.

2. *The learning procedure should be faster.*

In classical writing, the reader first read and then contemplate. In this approach, he is in a way forced (in a kind and respectful manner) to contemplate while reading.

3. *With graphical means, as lines, colours, spatial distribution of fact, we may help readers to easier memorize the subject.*

For instance, when we have four names of scholars, and to a name we associate the graphic of original signature, it will be easier to memorize, that there were two scholars, a scholar with round letters and another with sharp letters in their signatures. This is the example we used with the names of John Stuart Mill and Willian Herschel at the corresponding pages. If these two are memorized in this way, it is supposed that it would be easier to memorize the other two.

4. *The readers will be easily differentiated to those that just want to memorize the matter without investing much time, and others that have an intention to learn deeper while reading, and each of these categories will have equal chances to meet its interests by selecting the way of reading as we expressed in the section How to use this book.*

The active participation in anticipation of implicit presented information, which is required by readers, will support this differentiation and force a deeper understanding, thus faster and deeper learning.

Chapter 1
From Logic to Mathematical Logic

Although methods of logic and were obviously present in many cultures, which all used some intricate systems of reasoning, it is commonly accepted that explicit analysis of the principles of reasoning were developed independently in China, India, and Greece. The later being the most influential to the systems of logic in the West. In particular, the Aristotelian logic was widely accepted in the western science and mathematics. Many scholars contributed to the development, which has been continued by Islamic and medieval European scholars. The mid-fourteenth century is considered to be the period of the most respectable achievements. The period that followed including the first three or four decades of nineteenth century is viewed as a barren period or even period of declination and degradation of logic.

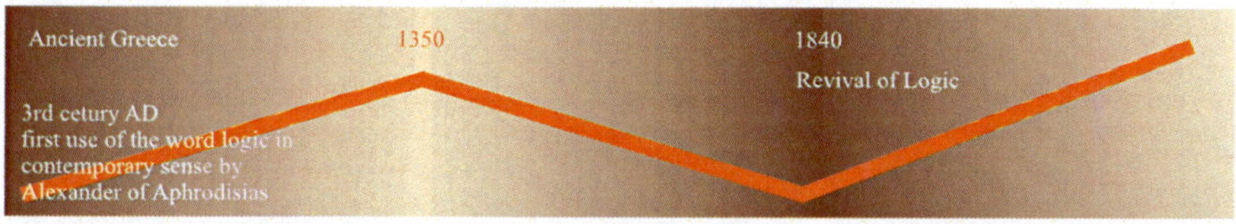

R.S. Stanković & J. Astola: From Boolean Log. to Switch. Circuits & Automata, SCI 335, pp. 3–33.
springerlink.com

1.1 Revival of Study of Logic in Britain

Richard Whately, 1826 John F.W. Herschel, 1830 William Whewell, 1837, 1840 John Stuart Mill, 1843	*Inductive Logic*
Augustus De Morgan, 1847 George J. Boole, 1847	*Symbolic logic*

Richard Whately

1.2 Elements of Logic by Whately

Anglican Archbishop of Dublin

1826, *Elements of Logic* an article in Encyclopaedia Metropolitana, has been reprinted as a separate publication in different variants several times.

The Encyclopaedia Metropolitana
was published in London, 1845,
quatro, 30 vols., and was issued
in 59 parts in 1817-1845
(22,426 pages, 565 plates).

George J. Boole, Wiliam Stanley Jevons, and many others referred to this particular work by R. Whately.

Reprinted from the ninth of the article by R. Whately (see page 17). So many editions clearly confirm the quality of the book.

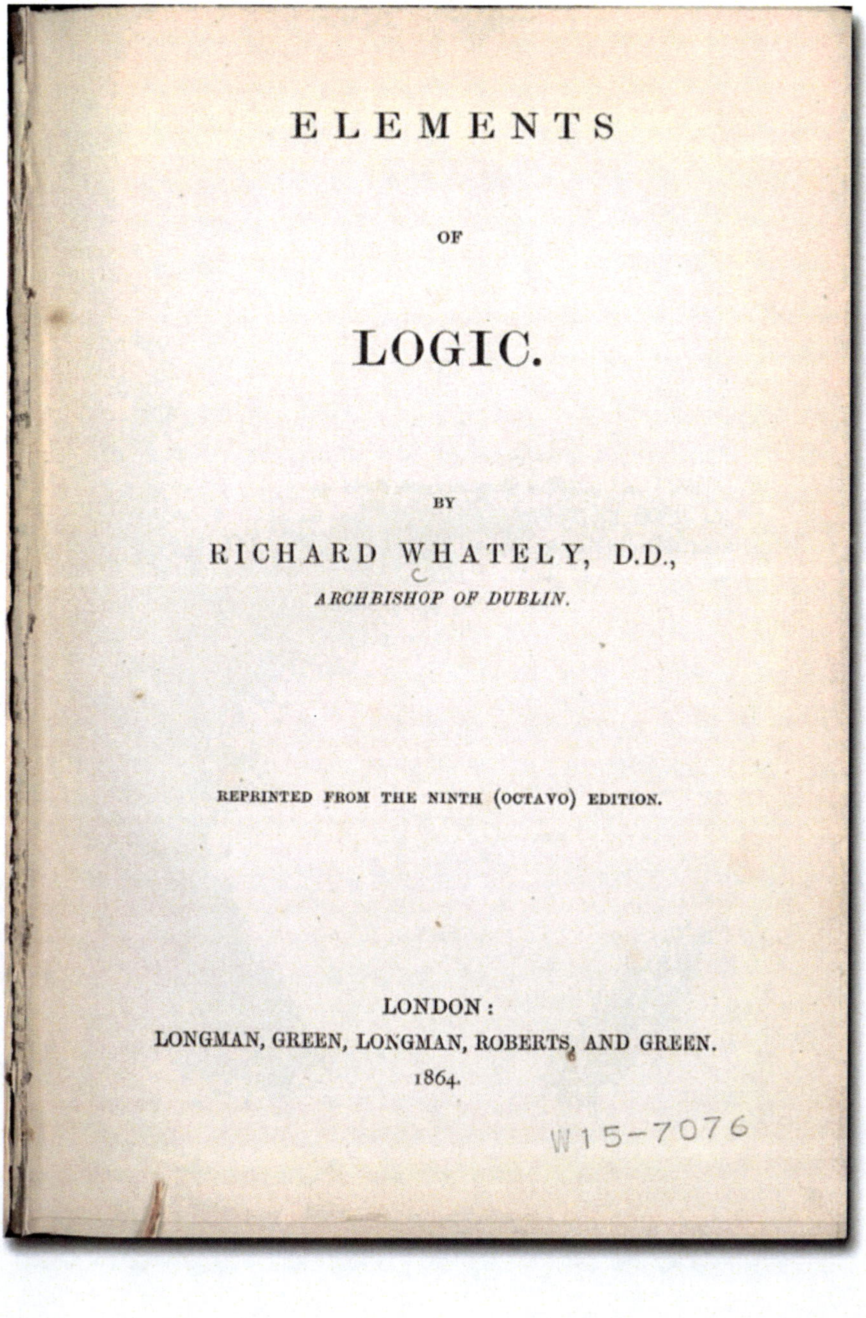

ELEMENTS

OF

LOGIC.

BY

RICHARD WHATELY, D.D.,

ARCHBISHOP OF DUBLIN.

REPRINTED FROM THE NINTH (OCTAVO) EDITION.

LONDON:

LONGMAN, GREEN, LONGMAN, ROBERTS, AND GREEN.

1864.

Analytical outline of Sciences, a discussion of reasoning as a universal system.

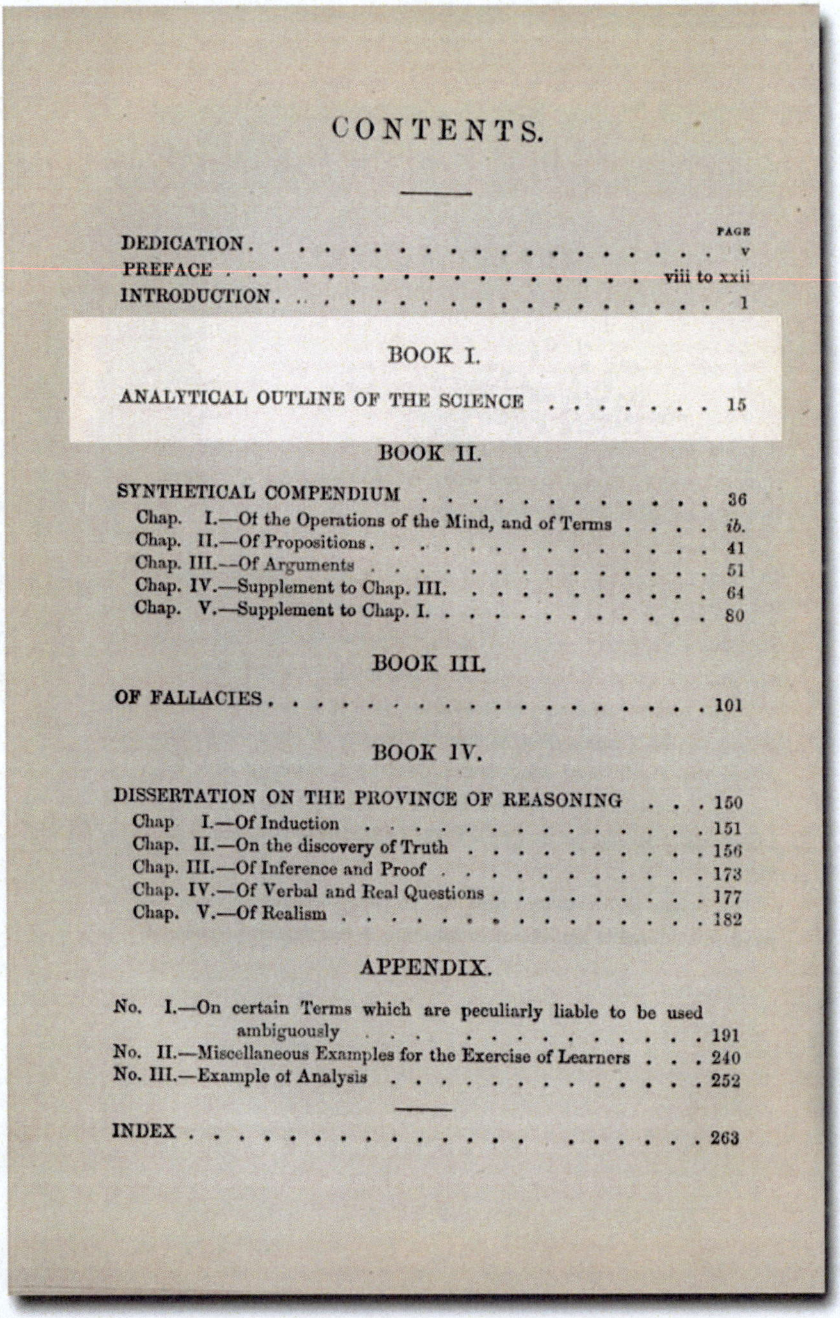

CONTENTS.

PAGE

DEDICATION v
PREFACE . . . . viii to xxii
INTRODUCTION 1

BOOK I.

ANALYTICAL OUTLINE OF THE SCIENCE 15

BOOK II.

SYNTHETICAL COMPENDIUM 36
Chap. I.—Of the Operations of the Mind, and of Terms . . . . *ib.*
Chap. II.—Of Propositions 41
Chap. III.—Of Arguments 51
Chap. IV.—Supplement to Chap. III. 64
Chap. V.—Supplement to Chap. I. 80

BOOK III.

OF FALLACIES 101

BOOK IV.

DISSERTATION ON THE PROVINCE OF REASONING . . . 150
Chap I.—Of Induction 151
Chap. II.—On the discovery of Truth 156
Chap. III.—Of Inference and Proof 173
Chap. IV.—Of Verbal and Real Questions 177
Chap. V.—Of Realism 182

APPENDIX.

No. I.—On certain Terms which are peculiarly liable to be used ambiguously 191
No. II.—Miscellaneous Examples for the Exercise of Learners . . . 240
No. III.—Example of Analysis 252

INDEX 263

In *Analytical Outline of the Science*, the first chapter of *Elements of Logic*, Whately pointed out that in every instance in which we do reasoning in the strict sense of this word, a certain process takes place in the mind. This process is unique in the sense that it is specific for this mental activity, it is based on certain rules and principles and it is the same in all cases of reasoning, provided that it will be correctly conducted.

The operation of reasoning is in all cases the same.

Whately said in*Elements of Logic*

It appear desirable to lay down some general rules of reasoning, applicable to all cases, by which a person might be enabled the more readily and clearly to state the grounds of his own conviction, or of his objection to the arguments of an opponent, instead of arguing at random, without any fixed and acknowledged principles to guide his procedure.

Such rules would be analogous to these of arithmetic, which obviate the tediousness and uncertainty of calculations in the head, wherein, after much of labours, different persons might arrive at different results, without any of them being able distinctly to point out the error of the rest.

Whately also discussed utility of Logic, and argued for its application as a general tool in reasoning.

Rev. Samuel Hinds, M.A., of Queen 's College and Vice -Principal of St. Alban Hall, Oxford, prepared a treatise *Introduction to Logic* comprising of 150 pages, *as a selection of the main theory in Elements of Logic by Whately, so much of that work as is necessary to qualify a candidate for examination in the Schools, and to that large class, whose only motive for studying logic is the attainm ent of this object, it is addressed*

Many editions and reprints from them confirm importance and strong influence of the work *Elements of Logic* by R. Whately.

Portability of information presentation media was an important issue already at that time.

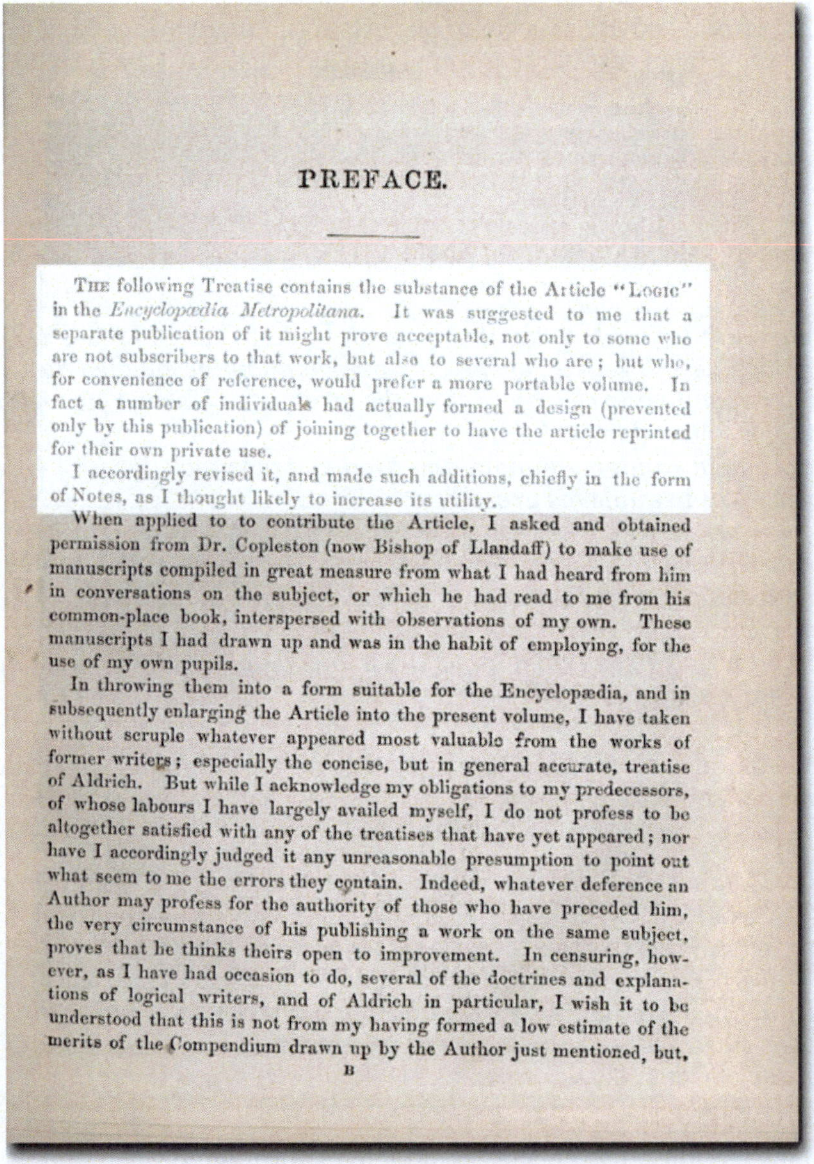

PREFACE.

THE following Treatise contains the substance of the Article "LOGIC" in the *Encyclopædia Metropolitana*. It was suggested to me that a separate publication of it might prove acceptable, not only to some who are not subscribers to that work, but also to several who are; but who, for convenience of reference, would prefer a more portable volume. In fact a number of individuals had actually formed a design (prevented only by this publication) of joining together to have the article reprinted for their own private use.

I accordingly revised it, and made such additions, chiefly in the form of Notes, as I thought likely to increase its utility.

When applied to to contribute the Article, I asked and obtained permission from Dr. Copleston (now Bishop of Llandaff) to make use of manuscripts compiled in great measure from what I had heard from him in conversations on the subject, or which he had read to me from his common-place book, interspersed with observations of my own. These manuscripts I had drawn up and was in the habit of employing, for the use of my own pupils.

In throwing them into a form suitable for the Encyclopædia, and in subsequently enlarging the Article into the present volume, I have taken without scruple whatever appeared most valuable from the works of former writers; especially the concise, but in general accurate, treatise of Aldrich. But while I acknowledge my obligations to my predecessors, of whose labours I have largely availed myself, I do not profess to be altogether satisfied with any of the treatises that have yet appeared; nor have I accordingly judged it any unreasonable presumption to point out what seem to me the errors they contain. Indeed, whatever deference an Author may profess for the authority of those who have preceded him, the very circumstance of his publishing a work on the same subject, proves that he thinks theirs open to improvement. In censuring, however, as I have had occasion to do, several of the doctrines and explanations of logical writers, and of Aldrich in particular, I wish it to be understood that this is not from my having formed a low estimate of the merits of the Compendium drawn up by the Author just mentioned, but,

B

A reference to the previous work by Dr. Copleston, which confirms that the subject has been intensively studied and something like lecture notes circulated among scholars and their pupils.

A discussion of utility of logic.

x PREFACE.

his aid composed, and by his permission inserted in the Encyclopædia, attracted so much attention as to occasion its separate publication, in a volume which has been frequently reprinted, not only in England, but in the United States of America; where it is in use, I believe, in every one of their Colleges. Add to which, the frequent allusions (compared with what could have been met with twenty or thirty years ago) to the subject of Logic, by writers on various subjects. And moreover several other treatises on the subject, either original works or abridgments, have been making their appearance with continually increased frequency of late years. Some indeed of these have little or nothing in common with the present work except the title. But even that very circumstance is so far encouraging, as indicating that the *name* of this science instead of exciting, as formerly, an almost universal prejudice, is considered as likely to prove a recommendation. Certainly Lord Dudley, were he *now* living, would not speak of the general neglect and contempt of Logic; though every branch of Science, Philosophy, and Literature, have flourished during the interval.

To explain fully the utility of Logic is what can be done only in the course of an explanation of the system itself. One preliminary observation only (for the original suggestion of which I am indebted to the same friend to whom this work is inscribed) it may be worth while to offer in this place. If it were inquired what is to be regarded as the most appropriate intellectual occupation of MAN, *as* man, what would be the answer? The Statesman is engaged with political affairs; the Soldier with military; the Mathematician, with the properties of numbers and magnitudes; the Merchant, with commercial concerns, &c.; but in what are *all* and each of these employed?—employed, I mean, as *men;* for there are many modes of exercise of the faculties, mental as well as bodily, which are in great measure common to us with the lower animals. Evidently, in *Reasoning.* They are all occupied in deducing, well or ill, Conclusions from Premises; each, concerning the subject of his own particular business. If, therefore, it be found that the process going on daily, in each of so many different minds, is, in any respect, the *same,* and if the principles on which it is conducted can be reduced to a regular system, and if rules can be deduced from that system, for the better conducting of the process, then, it can hardly be denied that such a system and such rules must be especially worthy the attention,—not of the members of this or that profession merely, but—of every one who is desirous of possessing a cultivated mind. To understand the theory of that which is the appropriate intellectual occupation of Man in general, and to learn to do that *well,* which every one will and *must* do, whether

Ninth edition of the article by R. Whately from Ecyclopaedia Metropolitana.

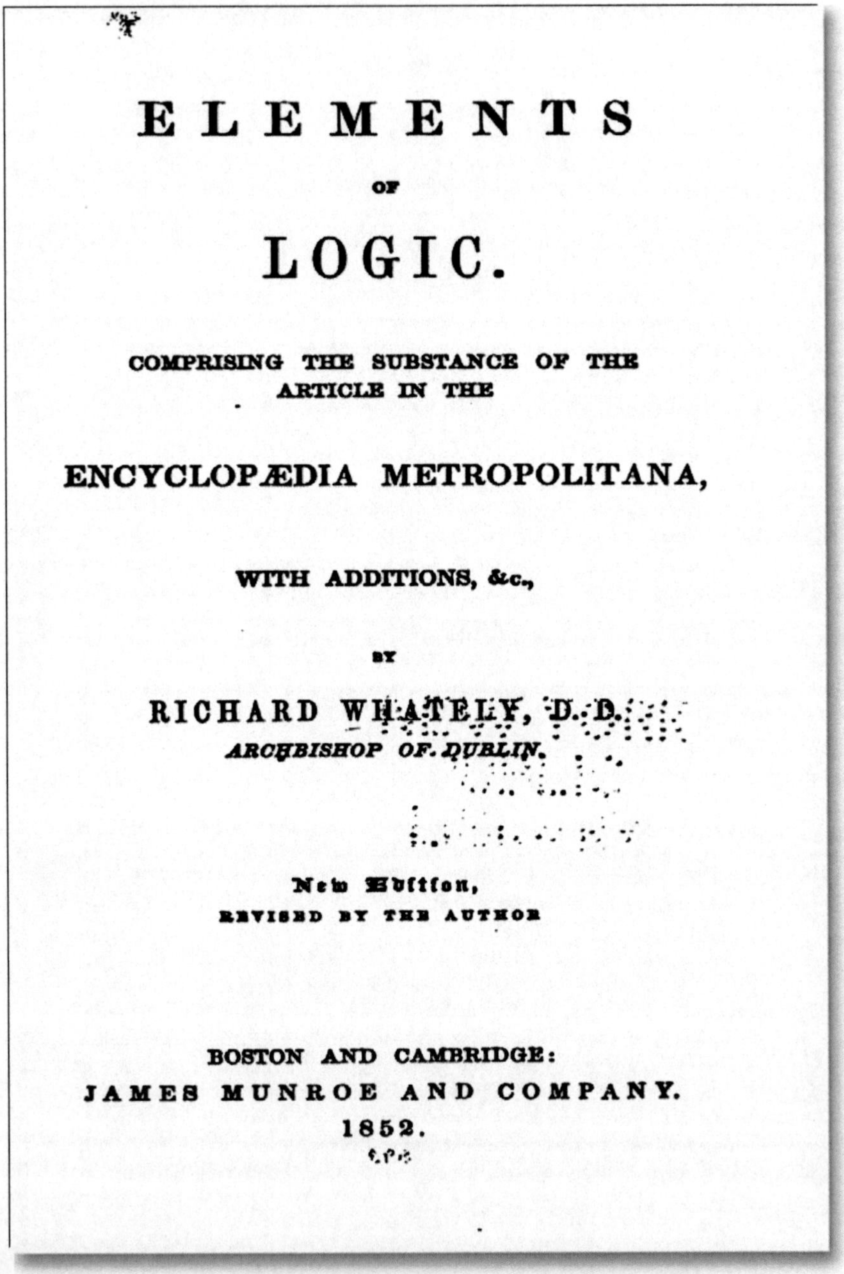

ELEMENTS

OF

LOGIC.

COMPRISING THE SUBSTANCE OF THE ARTICLE IN THE

ENCYCLOPÆDIA METROPOLITANA,

WITH ADDITIONS, &c.,

BY

RICHARD WHATELY, D.D.

ARCHBISHOP OF DUBLIN.

New Edition,

REVISED BY THE AUTHOR

BOSTON AND CAMBRIDGE:

JAMES MUNROE AND COMPANY.

1852.

John Frederick William Herschel

1.3 Ten Rules of Philosophizing by Herschel

J.F.W. Herschel used to publish in

Encyclopaedia Metropolitana

Cabinet Cyclopaedia

Edinburgh Encyclopaedia

In *A Preliminary Discourse on the Study of Natural Philosophy*, Cabinet Cyclopaedia, London, Series 1, Vol. 12, 1830, Herschel discusses rules of induction and hypotetico-deductive method with concrete examples from physics.

This treatise contains *Ten rules of philosophizing* intended for discovering and confirming causal relations in explaining some phenomena.

For Herschel, the Nature is governed by laws which are difficult to discern or to express by mathematical means, and the highest aim of natural philosophy is to understand these laws through inductive reasoning the result of which is to find a*single unifying explanation for a phenomenon.*

This work has a considerable influence to Charles Darwin, James Clerk Maxwell and John Stuart Mill.

John Herschel, son of the famous astronomer, William Herschel, the discoverer of Uranus, worked in the same area and discovered 3347 double stars and 2400 nebulae. He played a key role in the debates about discovery of Neptune around 1840, and related letters express his eloquent command of the English language.

Wiliam Herschel

Herschel, J. F. W., *Outlines of Astronomy*, in *Treasury of World Science.* Editor Dagobert D. Runes, (New York, *Philosophical Library*, 1962) , 449-458.

The highest aim of natural philosophy is to understand the laws of Nature through inductive reasoning.

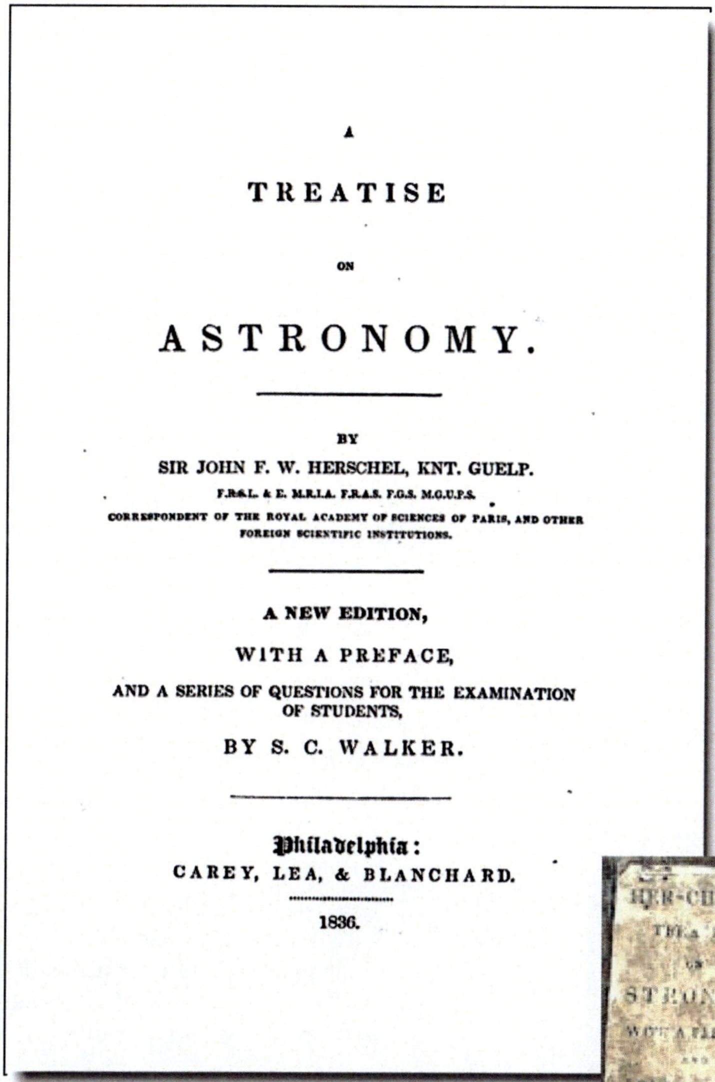

A

TREATISE

ON

ASTRONOMY.

BY

SIR JOHN F. W. HERSCHEL, KNT. GUELP.

F.R.S.L. & E. M.R.I.A. F.R.A.S. F.G.S. M.C.U.P.S.

CORRESPONDENT OF THE ROYAL ACADEMY OF SCIENCES OF PARIS, AND OTHER FOREIGN SCIENTIFIC INSTITUTIONS.

A NEW EDITION,

WITH A PREFACE,

AND A SERIES OF QUESTIONS FOR THE EXAMINATION OF STUDENTS,

BY S. C. WALKER.

Philadelphia:

CAREY, LEA, & BLANCHARD.

1836.

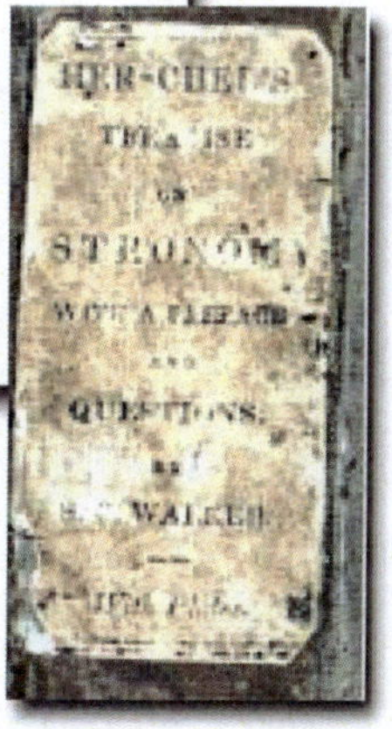

First page and a part of the cover.

John Herschel, detail of pencil drawing
by H.W. Pickersgill,
in the National Portrait Gallery, London

William Whewell

1.4 Inductive Science by Whewell

Important works relevant to the subject of Logic

History of the Inductive Sciences,
from Earliest to the Present Time, 1837

The Philosophy of the Inductive Sciences,
Founded Upon Their History, 1840

Whewell had a particular skill for devising new terms, he coined the words *anode*, *chatode*, *ion* for Faraday and the word *scientist* in 1833

William Whewell provided a classification of sciences.

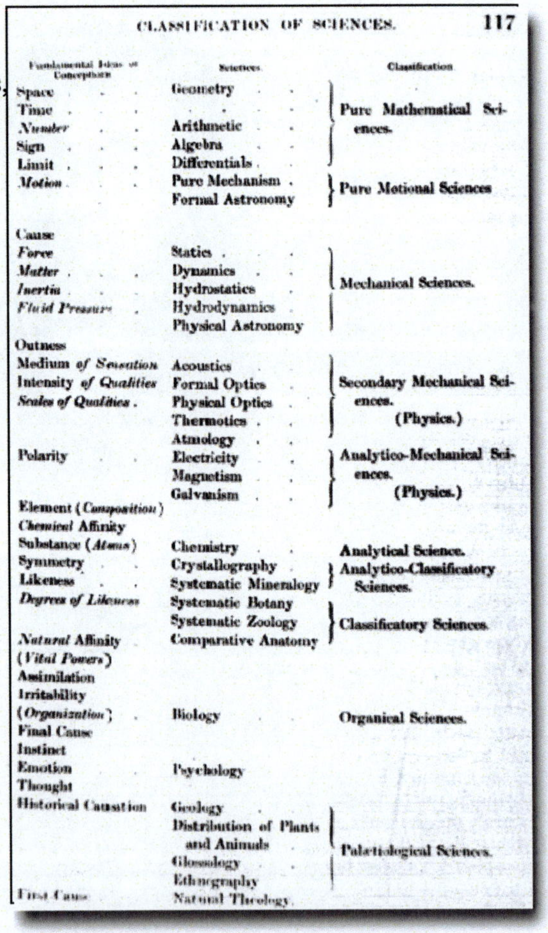
CLASSIFICATION OF SCIENCES. 117

Fundamental Ideas or Conceptions	Sciences	Classification
Space	Geometry	Pure Mathematical Sciences.
Time		
Number	Arithmetic	
Sign	Algebra	
Limit	Differentials	
Motion	Pure Mechanism	Pure Motional Sciences
	Formal Astronomy	
Cause		
Force	Statics	Mechanical Sciences.
Matter	Dynamics	
Inertia	Hydrostatics	
Fluid Pressure	Hydrodynamics	
	Physical Astronomy	
Outness		
Medium *of Sensation*	Acoustics	Secondary Mechanical Sciences. (Physics.)
Intensity *of Qualities*	Formal Optics	
Scales of Qualities	Physical Optics	
	Thermotics	
	Atmology	
Polarity	Electricity	Analytico-Mechanical Sciences. (Physics.)
	Magnetism	
	Galvanism	
Element (*Composition*)		
Chemical Affinity		
Substance (*Atoms*)	Chemistry	Analytical Science.
Symmetry	Crystallography	Analytico-Classificatory Sciences.
Likeness	Systematic Mineralogy	
Degrees of Likeness	Systematic Botany	Classificatory Sciences.
	Systematic Zoology	
Natural Affinity	Comparative Anatomy	
(*Vital Powers*)		
Assimilation		
Irritability		
(*Organization*)	Biology	Organical Sciences.
Final Cause		
Instinct		
Emotion	Psychology	
Thought		
Historical Causation	Geology	Palætiological Sciences.
	Distribution of Plants and Animals	
	Glossology	
	Ethnography	
First Cause	Natural Theology.	

Studied, and tutored at Trinity College, Cambridge

Worked jointly with *George Peacock* and *John Herschel* towards reforming the method of teaching the mathematics in Cambridge. These ideas are presented in *An Elementary Treatise on Mechanics* in 1819.

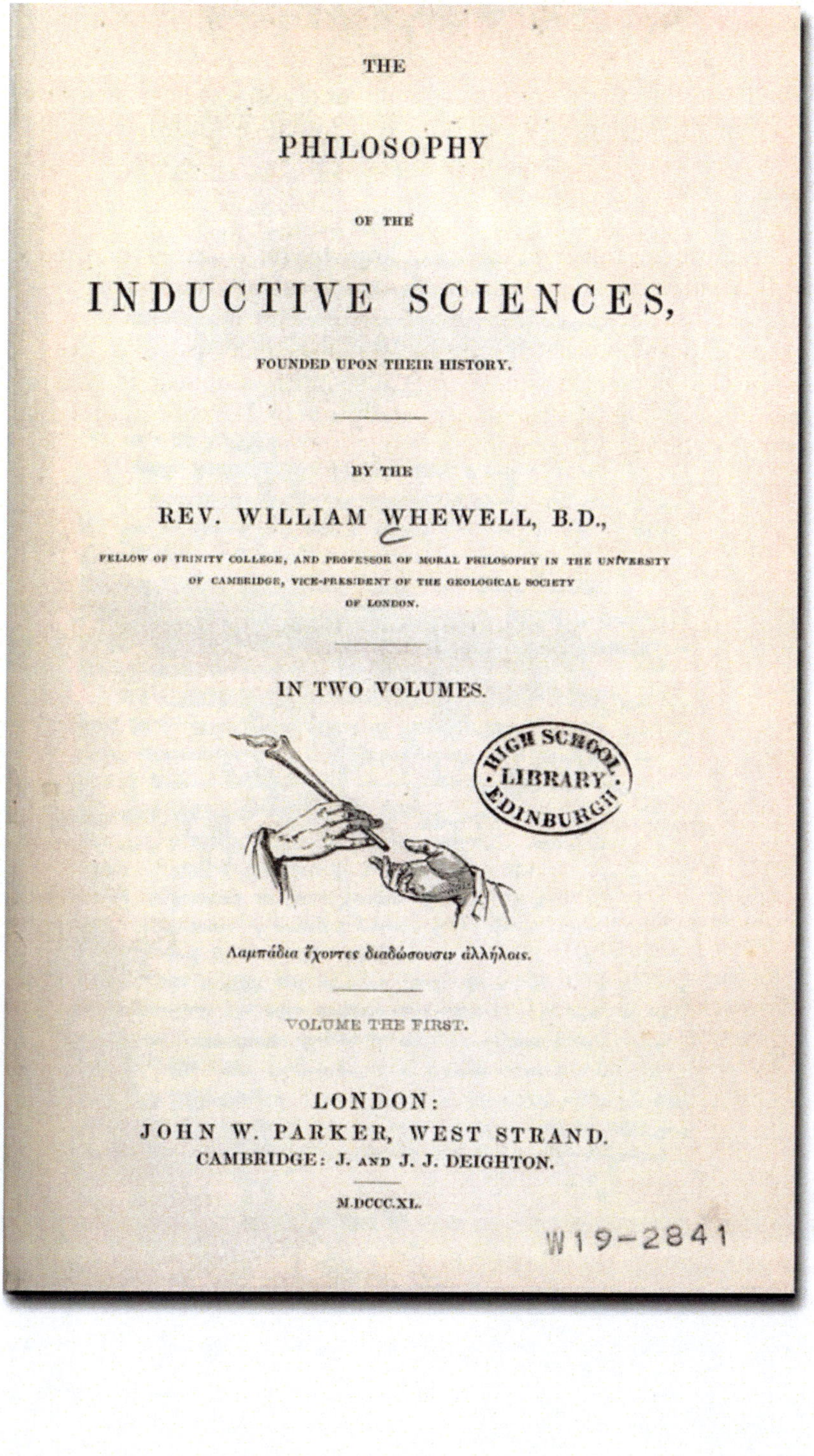

THE

PHILOSOPHY

OF THE

INDUCTIVE SCIENCES,

FOUNDED UPON THEIR HISTORY.

BY THE

REV. WILLIAM WHEWELL, B.D.,

FELLOW OF TRINITY COLLEGE, AND PROFESSOR OF MORAL PHILOSOPHY IN THE UNIVERSITY OF CAMBRIDGE, VICE-PRESIDENT OF THE GEOLOGICAL SOCIETY OF LONDON.

IN TWO VOLUMES.

Λαμπάδια ἔχοντες διαδώσουσιν ἀλλήλοις.

VOLUME THE FIRST.

LONDON:
JOHN W. PARKER, WEST STRAND.
CAMBRIDGE: J. AND J. J. DEIGHTON.

M.DCCC.XL.

Whewell is widely regarded as the father of modern philosophy of science.

He believed that there are fundamental and uniform laws of science, and knowing these laws we are able to acquire the knowledge of the world. These laws are exactly the Ideas that God used in creating the World.

In extracting an universal theory of knowledge, Whewell followed a middle road between pure rationalism and ultra-empiricism.

However, these believes of links betwen laws of science and ideas used by God in creating the World caused him to oppose the idea of evolution and later the theories of evolution by Darwin. In this case, he did not followed the middle but an extreme road.

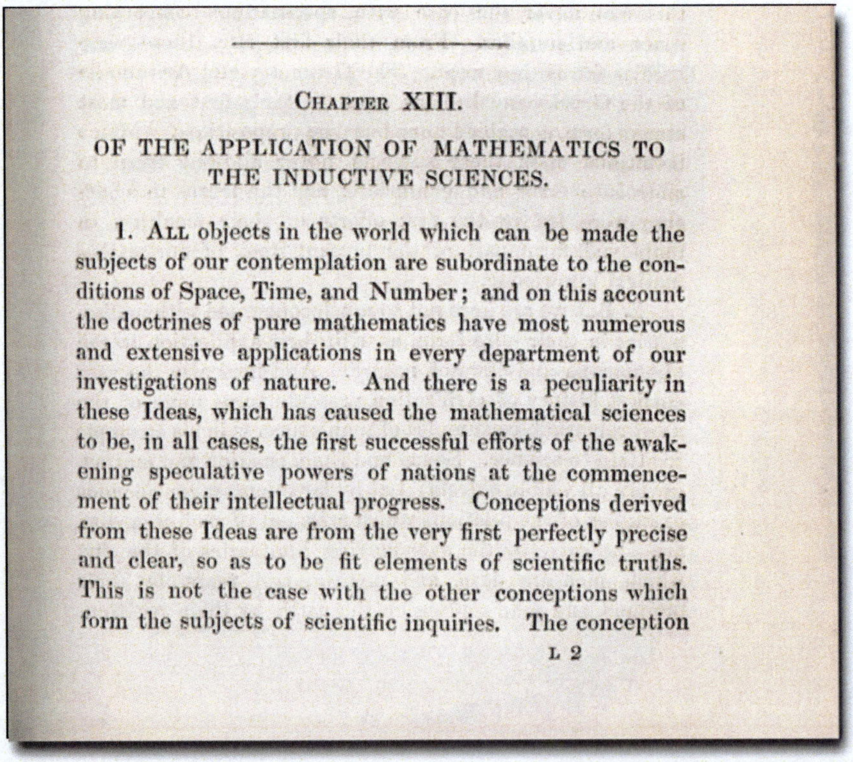

CHAPTER XIII.

OF THE APPLICATION OF MATHEMATICS TO THE INDUCTIVE SCIENCES.

1. ALL objects in the world which can be made the subjects of our contemplation are subordinate to the conditions of Space, Time, and Number; and on this account the doctrines of pure mathematics have most numerous and extensive applications in every department of our investigations of nature. And there is a peculiarity in these Ideas, which has caused the mathematical sciences to be, in all cases, the first successful efforts of the awakening speculative powers of nations at the commencement of their intellectual progress. Conceptions derived from these Ideas are from the very first perfectly precise and clear, so as to be fit elements of scientific truths. This is not the case with the other conceptions which form the subjects of scientific inquiries. The conception

L 2

Application of mathematical methods to inductive reasoning.

Induction as a method for disclosing laws of Nature.

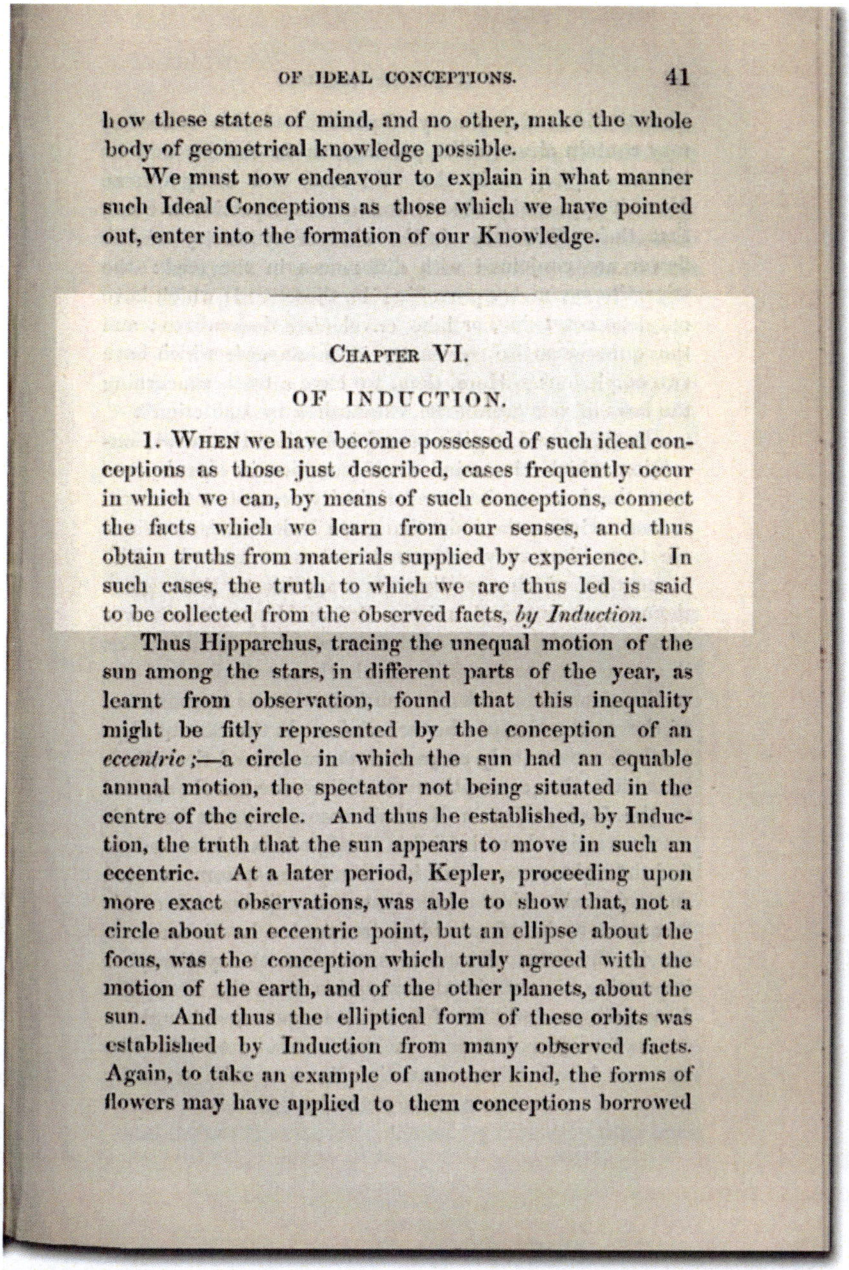

how these states of mind, and no other, make the whole body of geometrical knowledge possible.

We must now endeavour to explain in what manner such Ideal Conceptions as those which we have pointed out, enter into the formation of our Knowledge.

Chapter VI.

OF INDUCTION.

1. When we have become possessed of such ideal conceptions as those just described, cases frequently occur in which we can, by means of such conceptions, connect the facts which we learn from our senses, and thus obtain truths from materials supplied by experience. In such cases, the truth to which we are thus led is said to be collected from the observed facts, *by Induction.*

Thus Hipparchus, tracing the unequal motion of the sun among the stars, in different parts of the year, as learnt from observation, found that this inequality might be fitly represented by the conception of an *eccentric*;—a circle in which the sun had an equable annual motion, the spectator not being situated in the centre of the circle. And thus he established, by Induction, the truth that the sun appears to move in such an eccentric. At a later period, Kepler, proceeding upon more exact observations, was able to show that, not a circle about an eccentric point, but an ellipse about the focus, was the conception which truly agreed with the motion of the earth, and of the other planets, about the sun. And thus the elliptical form of these orbits was established by Induction from many observed facts. Again, to take an example of another kind, the forms of flowers may have applied to them conceptions borrowed

For Whewell, there are *facts* (data) that we can observe and which can be expressed by *ideas*. Examples of ideas are space, number, cause, likeness (resemblance) , etc.

The ideas are a special class of *conceptions*. Examples of conceptions are circle, a uniform force, etc.

The fact expressed by ideas have to be connected or unified by a mental act adding conceptions. Thus, the conceptions are used to *colligate* the facts. To simplify, if no conception added, then no real induction, since for Whewell, colligation is an essential feature of any induction, and in this way it is also essential to what Whewell calls *consilience* of inductions.

This understanding of induction has been disputed by some other scholars, in partiuclar these being more empiric oriented, as for instance John Stuart Mill.

In *Philosophy of the Inductive Sciences*, Whewell wrote

The Consilience of Inductions takes place when an Induction, obtained from one class of facts, coincides with an Induction obtained from another different class. Thus Consilience is a test of the truth of the Theory in which it occurs.

Whewel consider that in every *act of knowledge* there are two opposite elements, which we may call *Ideas* and *Perceptions*.

Each science has a *Particular Fundamental Idea* upon which facts that science considers are organized. Space, Cause, and substance, are fundamental ideas in Geometry, Mechanics, and Chemistry, respectively.

Each *Fundamental Idea* involves certain *conceptions*, which are *special modifications* of the Idea depending on concrete circumstances. An example of conception is force which is a modification of the Idea of Cause, when related to the particular case of motion.

John Stuart Mill

Deductive logic is the way of reasoning where we construct or evaluate deductive arguments which can be either valid or invalid. Valid arguments produce conclusions that are a logical consequence of the premises. If premises are true, then the conclusion is also true. We say that in a valid deductive argument the premises logically entail the conclusion, where the term entailment means that the truth of the premises is a guarantee of the truth of the conclusion.

Inductive logic extends the reasoning to cases where there are less than certain inferences, and the idea is to make conclusions and hypotheses based on observation of a certain number of instances.

Therefore, in a good inductive argument, the premises provide so me degree of support for the conclusions, and do not entail it, meaning that the premises do not ensure the truth of conclusions. As evidences accumulates, the degree to which the collection of true evidences supports a hypothesis should tend to indicate that the false hypothesis are false and true hypothesis are probably true.

Since in attempting to discover secrets of the Nature we are always restricted to the observation of a finite number of instances, the inductive logic is the right way to disclose and formulate laws describing natural phenomena.

In 1843 John Stuart Mill presented in the book *A System of Logic, Ratiocinative and Inductive* five principles of inductive reasoning called the Mills methods.

1.5 Mill and a System of Logic

J.S. Mill, *A System of Logic*, 1843

Induction, Mills method

Canons of Induction by Mill are viewed as an elaboration and reformulation of some considerations of John Herchsel.

Ten rules of philosophizing by Herschel are reformulated into *Five Canons of Induction*

1. Method of agreement
2. Method of difference
3. Joint method of agreement and difference
4. Method of residues
5. Method of concomitant variations

Canons eliminate hypotheses incompatible with given data and the law of universal causalisation states that there is a cause for every event.

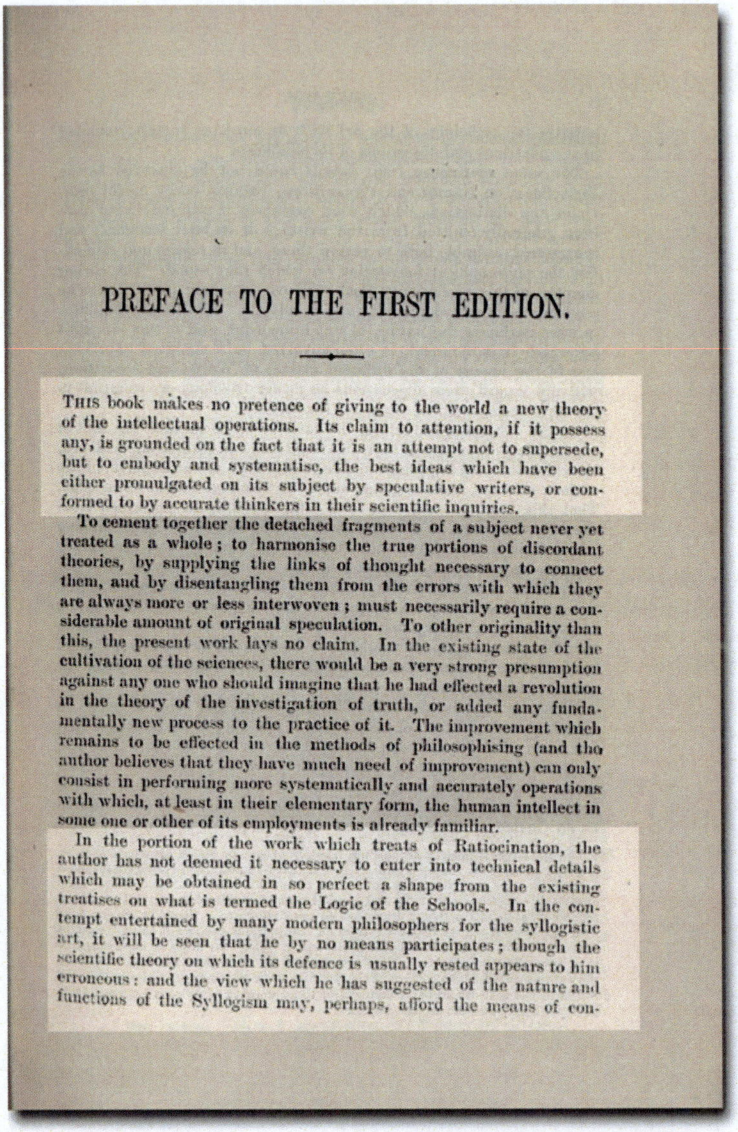

PREFACE TO THE FIRST EDITION.

THIS book makes no pretence of giving to the world a new theory of the intellectual operations. Its claim to attention, if it possess any, is grounded on the fact that it is an attempt not to supersede, but to embody and systematise, the best ideas which have been either promulgated on its subject by speculative writers, or conformed to by accurate thinkers in their scientific inquiries.

To cement together the detached fragments of a subject never yet treated as a whole; to harmonise the true portions of discordant theories, by supplying the links of thought necessary to connect them, and by disentangling them from the errors with which they are always more or less interwoven; must necessarily require a considerable amount of original speculation. To other originality than this, the present work lays no claim. In the existing state of the cultivation of the sciences, there would be a very strong presumption against any one who should imagine that he had effected a revolution in the theory of the investigation of truth, or added any fundamentally new process to the practice of it. The improvement which remains to be effected in the methods of philosophising (and the author believes that they have much need of improvement) can only consist in performing more systematically and accurately operations with which, at least in their elementary form, the human intellect in some one or other of its employments is already familiar.

In the portion of the work which treats of Ratiocination, the author has not deemed it necessary to enter into technical details which may be obtained in so perfect a shape from the existing treatises on what is termed the Logic of the Schools. In the contempt entertained by many modern philosophers for the syllogistic art, it will be seen that he by no means participates; though the scientific theory on which its defence is usually rested appears to him erroneous: and the view which he has suggested of the nature and functions of the Syllogism may, perhaps, afford the means of con-

Important here is the usage of the word *Ratiocination* - a reasoned train of thinking, process of exact thinking.

Very careful and detailed comments about related work of other authors and referencing to them. Important for the professionalism of scientists and writing nowadays.

Induction as a method and a reference to Whewell.

ciliating the principles of the art with as much as is well grounded in the doctrines and objections of its assailants.

The same abstinence from details could not be observed in the First Book, on Names and Propositions, because many useful principles and distinctions which were contained in the old Logic have been gradually omitted from the writings of its later teachers; and it appeared desirable both to revive these, and to reform and rationalise the philosophical foundation on which they stood. The earlier chapters of this preliminary Book will consequently appear, to some readers, needlessly elementary and scholastic. But those who know in what darkness the nature of our knowledge, and of the processes by which it is obtained, is often involved by a confused apprehension of the import of the different classes of Words and Assertions, will not regard these discussions as either frivolous or irrelevant to the topics considered in the latter Books.

On the subject of Induction, the task to be performed was that of generalising the modes of investigating truth and estimating evidence, by which so many important and recondite laws of nature have, in the various sciences, been aggregated to the stock of human knowledge. That this is not a task free from difficulty may be presumed from the fact, that even at a very recent period, eminent writers (among whom it is sufficient to name Archbishop Whately, and the author of a celebrated article on Bacon in the *Edinburgh Review*) have not scrupled to pronounce it impossible.* The author has endeavoured to combat their theory in the manner in which Diogenes confuted the sceptical reasonings against the possibility of motion; remembering that Diogenes' argument would have been equally conclusive, though his individual perambulations might not have extended beyond the circuit of his own tub.

Whatever may be the value of what the author has succeeded in effecting on this branch of his subject, it is a duty to acknowledge that for much of it he has been indebted to several important treatises, partly historical and partly philosophical, on the generalities and processes of physical science, which have been published within the last few years. To these treatises, and to their authors, he has endeavoured to do justice in the body of the work. But as with one of these

* In the later editions of Archbishop Whately's *Logic*, he states his meaning to be, not that "rules" for the ascertainment of truths by inductive investigation cannot be laid down, or that they may not be "of eminent service," but that they "must always be comparatively vague and general, and incapable of being built up into a regular demonstrative theory like that of the Syllogism." (Book iv. ch. iv. § 3.) And he observes, that to devise a system for this purpose, capable of being "brought into a scientific form," would be an achievement which "he must be more sanguine than scientific who expects." (Book iv. ch. ii. § 4.) To effect this, however, being the express object of the portion of the present work which treats of Induction, the words in the text are no overstatement of the difference of opinion between Archbishop Whately and me on the subject.

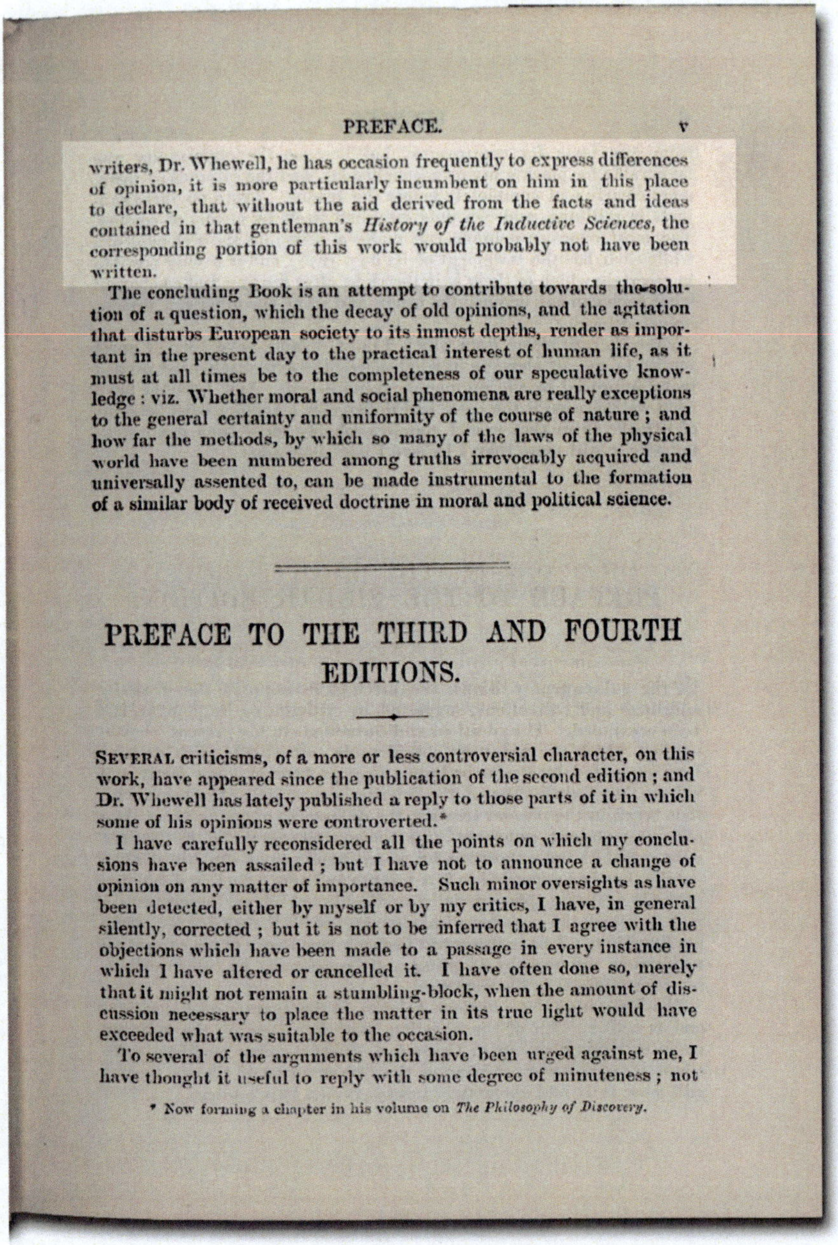

writers, Dr. Whewell, he has occasion frequently to express differences of opinion, it is more particularly incumbent on him in this place to declare, that without the aid derived from the facts and ideas contained in that gentleman's *History of the Inductive Sciences,* the corresponding portion of this work would probably not have been written.

The concluding Book is an attempt to contribute towards the solution of a question, which the decay of old opinions, and the agitation that disturbs European society to its inmost depths, render as important in the present day to the practical interest of human life, as it must at all times be to the completeness of our speculative knowledge: viz. Whether moral and social phenomena are really exceptions to the general certainty and uniformity of the course of nature; and how far the methods, by which so many of the laws of the physical world have been numbered among truths irrevocably acquired and universally assented to, can be made instrumental to the formation of a similar body of received doctrine in moral and political science.

PREFACE TO THE THIRD AND FOURTH EDITIONS.

SEVERAL criticisms, of a more or less controversial character, on this work, have appeared since the publication of the second edition; and Dr. Whewell has lately published a reply to those parts of it in which some of his opinions were controverted.*

I have carefully reconsidered all the points on which my conclusions have been assailed; but I have not to announce a change of opinion on any matter of importance. Such minor oversights as have been detected, either by myself or by my critics, I have, in general silently, corrected; but it is not to be inferred that I agree with the objections which have been made to a passage in every instance in which I have altered or cancelled it. I have often done so, merely that it might not remain a stumbling-block, when the amount of discussion necessary to place the matter in its true light would have exceeded what was suitable to the occasion.

To several of the arguments which have been urged against me, I have thought it useful to reply with some degree of minuteness; not

* Now forming a chapter in his volume on *The Philosophy of Discovery.*

J.S.Mill wrote in *A System of Logic* that his work has been influenced by Whewell and his work *History of Inductuve Sciences*.

> has succeeded in effecting on this branch of his subject, it is a duty to acknowledge that for much of it he has been indebted to several important treatises, partly historical and partly philosophical, on the generalities and processes of physical science, which have been published within the last few years. To these treatises, and to their authors, he has endeavoured to do justice in the body of the work. But as with one of these writers, Dr. Whewell, he has occasion frequently to express differences of opinion, it is more particularly incumbent on him in this place to declare, that without the aid derived from the facts and ideas contained in that gentleman's *History of the Inductive Sciences*, the corresponding portion of this work would probably not have been written.

There are essential differences in understanding the inductive reasoning between Whewell and Mill. For instance, J.S. Mill said *experience shows that a triple of things can be divided into a double and a single.*

Whewell would say that because it is inconceivable to think that $2+1$ is not equal to 3, therefore $2+1=3$. Although the empirical evidence confirms this natural intuition, it is not the cause of it.

Reference to Whewell, important to notice since it is related to discussions and disagreement between Mill and Whewell, characterized by scholars as lovely debates.

Impressive ninth edition, extended and detailed comments on the following page.

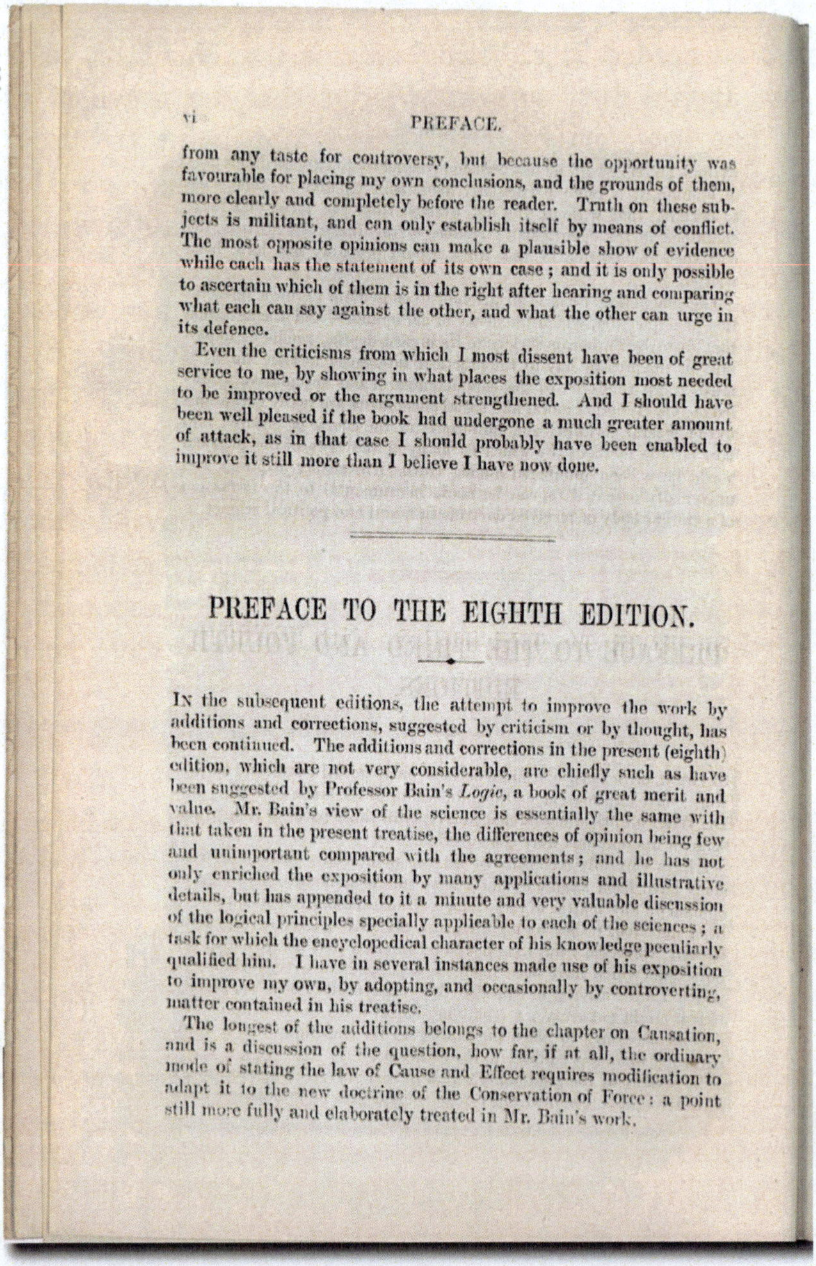

vi PREFACE.

from any taste for controversy, but because the opportunity was favourable for placing my own conclusions, and the grounds of them, more clearly and completely before the reader. Truth on these subjects is militant, and can only establish itself by means of conflict. The most opposite opinions can make a plausible show of evidence while each has the statement of its own case; and it is only possible to ascertain which of them is in the right after hearing and comparing what each can say against the other, and what the other can urge in its defence.

Even the criticisms from which I most dissent have been of great service to me, by showing in what places the exposition most needed to be improved or the argument strengthened. And I should have been well pleased if the book had undergone a much greater amount of attack, as in that case I should probably have been enabled to improve it still more than I believe I have now done.

PREFACE TO THE EIGHTH EDITION.

In the subsequent editions, the attempt to improve the work by additions and corrections, suggested by criticism or by thought, has been continued. The additions and corrections in the present (eighth) edition, which are not very considerable, are chiefly such as have been suggested by Professor Bain's *Logic*, a book of great merit and value. Mr. Bain's view of the science is essentially the same with that taken in the present treatise, the differences of opinion being few and unimportant compared with the agreements; and he has not only enriched the exposition by many applications and illustrative details, but has appended to it a minute and very valuable discussion of the logical principles specially applicable to each of the sciences; a task for which the encyclopedical character of his knowledge peculiarly qualified him. I have in several instances made use of his exposition to improve my own, by adopting, and occasionally by controverting, matter contained in his treatise.

The longest of the additions belongs to the chapter on Causation, and is a discussion of the question, how far, if at all, the ordinary mode of stating the law of Cause and Effect requires modification to adapt it to the new doctrine of the Conservation of Force: a point still more fully and elaborately treated in Mr. Bain's work.

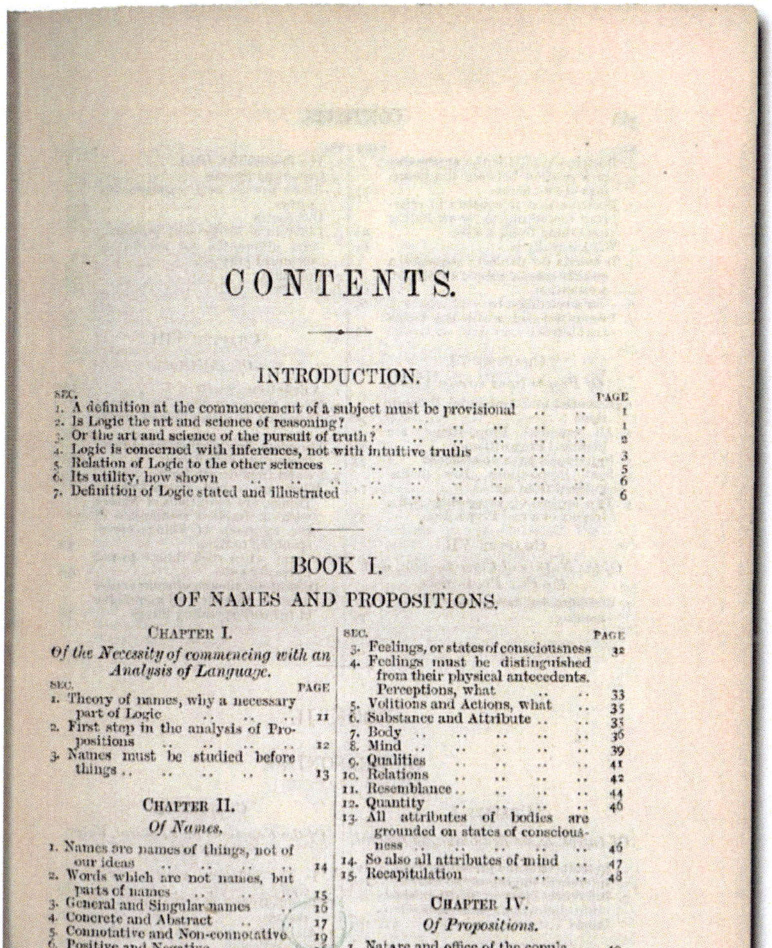

CONTENTS.

INTRODUCTION.

SEC. PAGE
1. A definition at the commencement of a subject must be provisional .. 1
2. Is Logic the art and science of reasoning? 1
3. Or the art and science of the pursuit of truth? 2
4. Logic is concerned with inferences, not with intuitive truths 3
5. Relation of Logic to the other sciences 5
6. Its utility, how shown 6
7. Definition of Logic stated and illustrated 6

BOOK I.

OF NAMES AND PROPOSITIONS.

Chapter I.

Of the Necessity of commencing with an Analysis of Language.

SEC. PAGE
1. Theory of names, why a necessary part of Logic 11
2. First step in the analysis of Propositions 12
3. Names must be studied before things 13

Chapter II.

Of Names.

1. Names are names of things, not of our ideas 14
2. Words which are not names, but parts of names 15
3. General and Singular names .. 16
4. Concrete and Abstract 17
5. Connotative and Non-connotative 19
6. Positive and Negative 26
7. Relative and Absolute 27
8. Univocal and Æquivocal 28

Chapter III.

Of the Things denoted by Names.

1. Necessity of an enumeration of Nameable Things. The Categories of Aristotle 29
2. Ambiguity of the most general names 30
3. Feelings, or states of consciousness 32
4. Feelings must be distinguished from their physical antecedents. Perceptions, what 33
5. Volitions and Actions, what .. 35
6. Substance and Attribute 35
7. Body 36
8. Mind 39
9. Qualities 41
10. Relations 42
11. Resemblance 44
12. Quantity 46
13. All attributes of bodies are grounded on states of consciousness 46
14. So also all attributes of mind .. 47
15. Recapitulation 48

Chapter IV.

Of Propositions.

1. Nat re and office of the copula .. 49
2. Affirmative and Negative propositions 51
3. Simple and Complex 52
4. Universal, Particular, and Singular 54

Chapter V.

Of the Import of Propositions.

1. Doctrine that a Proposition is the expression of a relation between two ideas 55

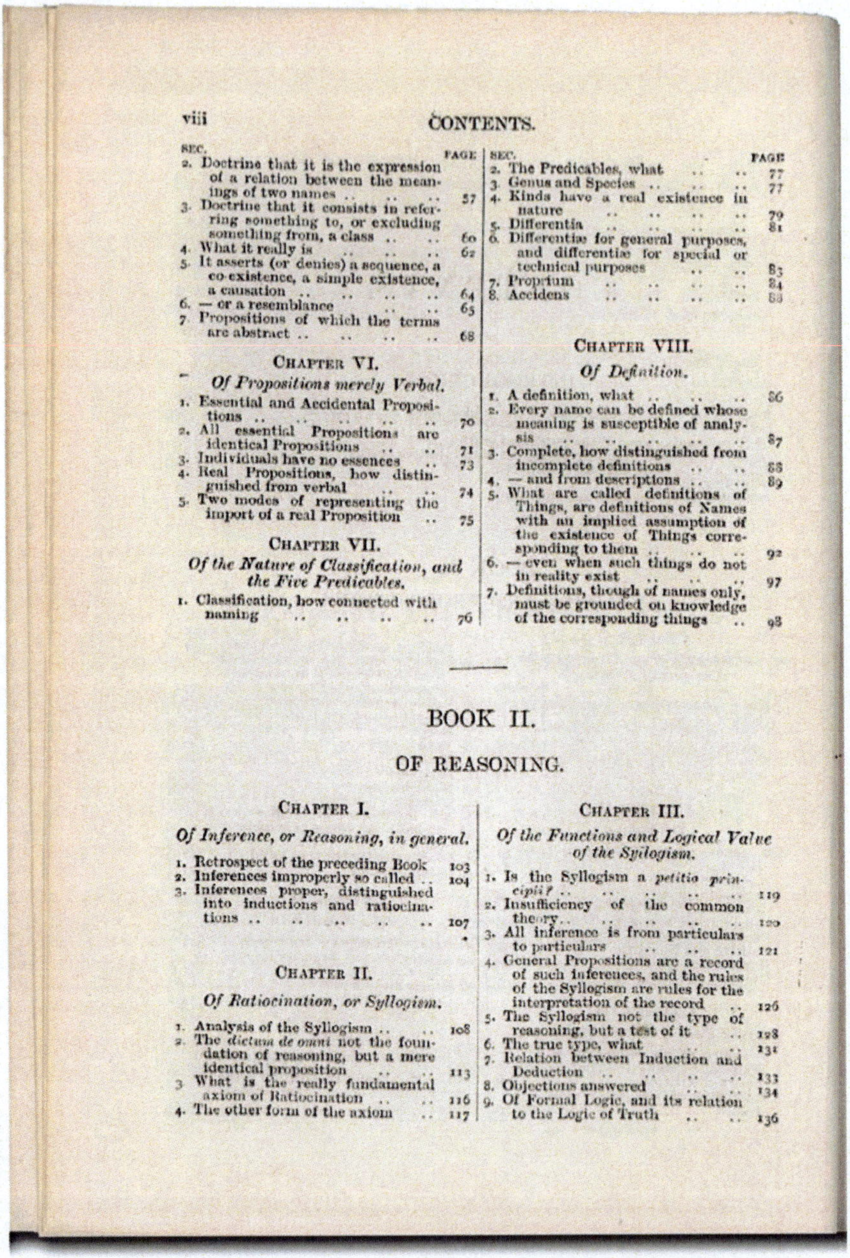

viii CONTENTS.

SEC. PAGE
2. Doctrine that it is the expression of a relation between the meanings of two names 57
3. Doctrine that it consists in referring something to, or excluding something from, a class 60
4. What it really is 62
5. It asserts (or denies) a sequence, a co-existence, a simple existence, a causation 64
6. — or a resemblance 65
7. Propositions of which the terms are abstract 68

CHAPTER VI.

Of Propositions merely Verbal.

1. Essential and Accidental Propositions 70
2. All essential Propositions are identical Propositions 71
3. Individuals have no essences .. 73
4. Real Propositions, how distinguished from verbal 74
5. Two modes of representing the import of a real Proposition .. 75

CHAPTER VII.

Of the Nature of Classification, and the Five Predicables.

1. Classification, how connected with naming 76
2. The Predicables, what 77
3. Genus and Species 77
4. Kinds have a real existence in nature 79
5. Differentia 81
6. Differentiæ for general purposes, and differentiæ for special or technical purposes 83
7. Proprium 84
8. Accidens 85

CHAPTER VIII.

Of Definition.

1. A definition, what 86
2. Every name can be defined whose meaning is susceptible of analysis 87
3. Complete, how distinguished from incomplete definitions 88
4. — and from descriptions 89
5. What are called definitions of Things, are definitions of Names with an implied assumption of the existence of Things corresponding to them 92
6. — even when such things do not in reality exist 97
7. Definitions, though of names only, must be grounded on knowledge of the corresponding things .. 98

BOOK II.

OF REASONING.

CHAPTER I.

Of Inference, or Reasoning, in general.

1. Retrospect of the preceding Book 103
2. Inferences improperly so called .. 104
3. Inferences proper, distinguished into inductions and ratiocinations 107

CHAPTER II.

Of Ratiocination, or Syllogism.

1. Analysis of the Syllogism 108
2. The *dictum de omni* not the foundation of reasoning, but a mere identical proposition 113
3. What is the really fundamental axiom of Ratiocination 116
4. The other form of the axiom .. 117

CHAPTER III.

Of the Functions and Logical Value of the Syllogism.

1. Is the Syllogism a *petitio principii?* 119
2. Insufficiency of the common theory 120
3. All inference is from particulars to particulars 121
4. General Propositions are a record of such inferences, and the rules of the Syllogism are rules for the interpretation of the record .. 126
5. The Syllogism not the type of reasoning, but a test of it .. 128
6. The true type, what 131
7. Relation between Induction and Deduction 133
8. Objections answered 134
9. Of Formal Logic, and its relation to the Logic of Truth 136

CONTENTS. ix

Chapter IV.

Of Trains of Reasoning, and Deductive Sciences.

SEC. PAGE

1. For what purpose trains of reasoning exist 137
2. A train of reasoning is a series of inductive inferences 138
3. — from particulars to particulars through marks of marks 139
4. Why there are deductive sciences 141
5. Why other sciences still remain experimental 143
6. Experimental sciences may become deductive by the progress of experiment 145
7. In what manner this usually takes place 145

Chapter V.

Of Demonstration, and Necessary Truths.

1. The Theorems of Geometry are necessary truths only in the sense of necessarily following from hypotheses 147
2. Those hypotheses are real facts with some of their circumstances exaggerated or omitted 149
3. Some of the first principles of Geometry are axioms, and these are not hypothetical 150
4. — but are experimental truths .. 151
5. An objection answered 153

SEC. PAGE

6. Dr. Whewell's opinions on axioms examined 156

Chapter VI.

The same Subject continued.

1. All deductive sciences are inductive 164
2. The propositions of the Science of Number are not verbal, but generalisations from experience .. 166
3. In what sense hypothetical .. 169
4. The characteristic property of demonstrative science is to be hypothetical 170
5. Definition of demonstrative evidence 171

Chapter VII.

Examination of some Opinions opposed to the preceding doctrines.

1. Doctrine of the Universal Postulate 172
2. The test of inconceivability does not represent the aggregate of past experience 173
3. — nor is implied in every process of thought 175
4. Objections answered 179
5. Sir W. Hamilton's opinion on the Principles of Contradiction and Excluded Middle 182

BOOK III.

OF INDUCTION.

Chapter I.

Preliminary Observations on Induction in General.

1. Importance of an Inductive Logic 185
2. The Logic of Science is also that of business and life 185

Chapter II.

Of Inductions improperly so called.

1. Inductions distinguished from verbal transformations 188
2. — from inductions, falsely so called, in mathematics 190
3. — and from descriptions 191
4. Examination of Dr. Whewell's theory of Induction 192
5. Further illustration of the preceding remarks 198

Chapter III.

On the Ground of Induction.

1. Axiom of the uniformity of the course of Nature 200
2. Not true in every sense. Induction *per enumerationem simplicem* 203
3. The question of Inductive Logic stated 205

Chapter IV.

Of Laws of Nature.

1. The general regularity in nature is a tissue of partial regularities, called laws 206
2. Scientific induction must be grounded on previous spontaneous inductions 208
3. Are there any inductions fitted to be a test of all others? 209

Chapter 2
Boolean Logic

R.S. Stanković & J. Astola: From Boolean Log. to Switch. Circuits & Automata, SCI 335, pp. 35–86.
springerlink.com

Augustus De Morgan

2.1 Formal Logic by De Morgan

De Morgan, A., *Formal Logic- or the calculus of inference, necessary and probable*, London 1847, xvi+336.

De Morgan Laws of duality

$$\overline{x \vee y} = \bar{x} \wedge \bar{y} \qquad \overline{x \wedge y} = \bar{x} \vee \bar{y}$$

Richard Whately

John F. W. Hercshel

William Whewell

John Stuart Mill

Intuitive Logic

George J. Boole

Augustus De Morgan

Symbolic Logic

De Morgan wrote reviews of the books by Boole and Jevons.

Sanchez Valencia, V.,"An Athenaeum curiosity De Morgan's reviews of Boole and Jevons", *History and Philosophy of Logic*, Vol. 22, No. 2, 2001, 75-79.

correspondence with Boole dated 1842-1864, about 90 letters

Smith, G.S. (ed.), *The Boole– De Morgan Correspondence*, Oxford University Press, 1982.

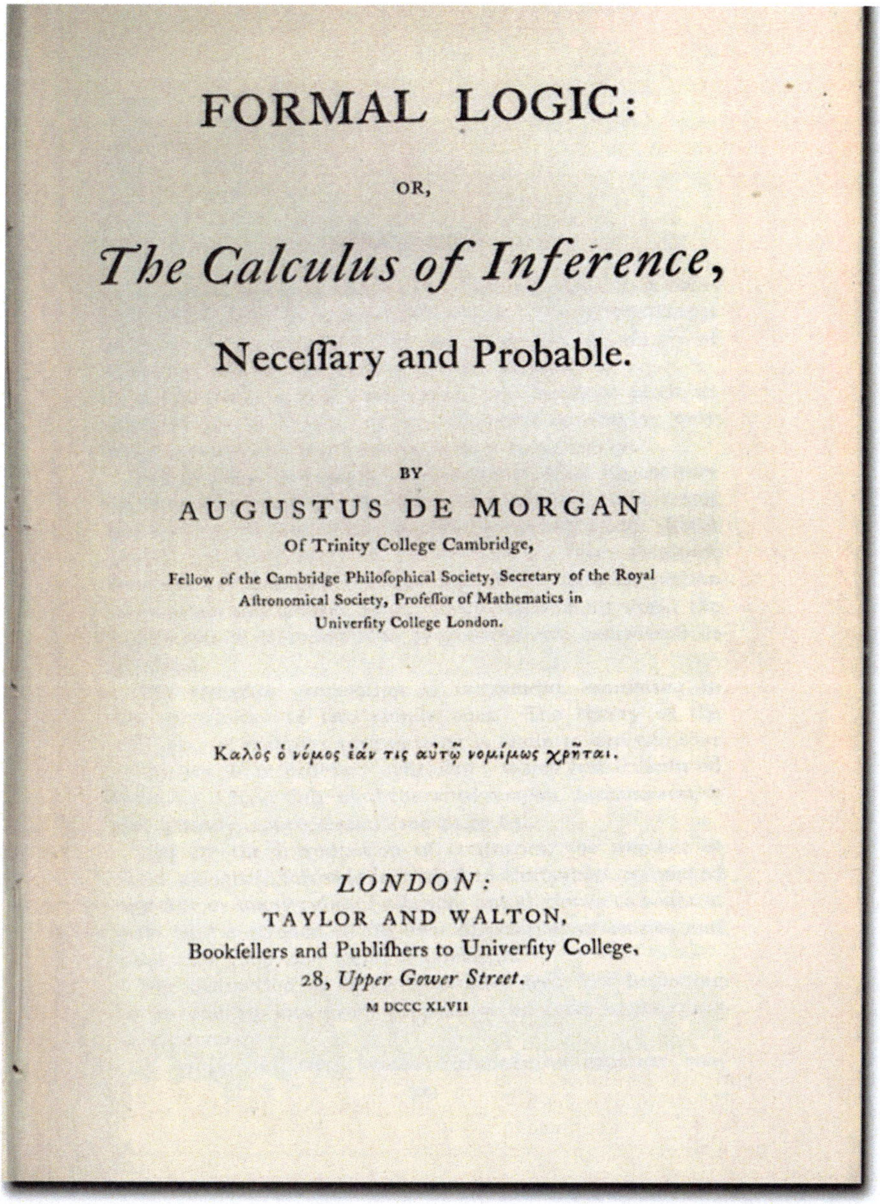

FORMAL LOGIC:

OR,

The Calculus of Inference,

Neceſſary and Probable.

BY

AUGUSTUS DE MORGAN

Of Trinity College Cambridge,

Fellow of the Cambridge Philoſophical Society, Secretary of the Royal Aſtronomical Society, Profeſſor of Mathematics in Univerſity College London.

Καλὸς ὁ νόμος ἐάν τις αὐτῷ νομίμως χρῆται.

LONDON:

TAYLOR AND WALTON,

Bookſellers and Publiſhers to Univerſity College,

28, *Upper Gower Street.*

M DCCC XLVII

FORMAL LOGIC
(1847)

BY
AUGUSTUS DE MORGAN

EDITED BY
A. E. TAYLOR, F.B.A.
PROFESSOR OF MORAL PHILOSOPHY AT
THE UNIVERSITY OF EDINBURGH

LONDON
THE OPEN COURT COMPANY
1926

Portrait of De Morgan from a pastel made by his wife. Reprinted in

For the second case we must write $m' = m + \xi - \eta = m + \eta' - \xi'$, and so on. I now proceed to the several divisions into which our usual modes of thinking make it convenient to separate the cases of this most general form.

[160] First, when every thing is numerically definite. In this case, as seen, every form requires an examination of the premises and conclusion, as to whether they are or are not spurious.

Secondly, when v, the number of instances in the whole universe of names, is wholly unknown. In this case ξ' is indefinite when ξ is definite, and *vice versâ*; and similarly one at least of each two, η or η', ζ or ζ', is indefinite. There are then no spurious conclusions; or, which is the same thing, none which are known to be such: for the spuriousness of a premise or conclusion consists in our *knowing* that it must be true of its two terms, independently of all comparison of those terms with a third.

Thirdly, when ξ, η, ζ, are all indefinite, as well as v. In this case, as here stated, there is no possibility of inference. We cannot tell whether $m+n$ be or be not greater than η, if we do not know what η is, in any manner, or to any extent.

But here we introduce that degree of definiteness by which we distinguish the *universal* from the *particular* (or *possible* particular, see page 56) proposition. If we can know that either of the two, m and n, is the same as η (greater neither can be) then we know that $m+n$ is greater than η. And at the same time we make Y universal, in one or the other of the premises. And the same if we can know that either m' or n' is η'.

The following are the forms which may all be derived from the first, by using all the varieties of contrary names and contranominal equivalents. If we want, for instance, to show the connection of the fourteenth with the first, we throw the first into the form

$$(m + \eta' - \xi)\mathrm{xy} + n\mathrm{YZ} = (m + n + \eta' - \xi - \zeta)\mathrm{xz}$$

THE NUMERICALLY DEFINITE SYLLOGISM 185

We then change x into X, and Z into z, changing at the same time ξ into ξ' and ζ into ζ': and thus we get

$$(m+\eta'-\xi')\mathrm{Xy}+n\mathrm{Yz}=(m+n+\eta'-\xi'-\zeta')\mathrm{XZ}$$

Now, for $m+\eta'-\xi'$ write m', that is, for m write $m'-\eta'+\xi'$ and we have

$$m'\mathrm{Xy}+n\mathrm{Yz}=(m'+n-\zeta')\mathrm{XZ}$$

which is one of the forms of the fourteenth. And $(n+m'-\xi)$ xz is only the contranominal of $(m'+n-\zeta')$XZ.

[161]

1. $m\mathrm{XY}\ +n\mathrm{YZ}=(m+n-\eta)\mathrm{XZ}=(m+n+\eta'-\xi-\zeta)\mathrm{xz}$
2. $m'\mathrm{xy}\ +n\mathrm{YZ}=(n+m'-\xi')\mathrm{XZ}=(m'+n-\zeta)\mathrm{xz}$
3. $m\mathrm{XY}\ +n'\mathrm{yz}=(m+n'-\zeta')\mathrm{XZ}=(n'+m-\xi)\mathrm{xz}$
4. $m'\mathrm{xy}\ +n'\mathrm{yz}=(m'+n'+\eta-\zeta'-\xi')\mathrm{XZ}=(m'+n'-\eta')\mathrm{xz}$
5. $m\mathrm{xY}\ +n\mathrm{YZ}=(m+n-\eta)\mathrm{xZ}=(m+n+\eta'-\xi'-\zeta)\mathrm{Xz}$
6. $m'\mathrm{Xy}\ +n\mathrm{YZ}=(n+m'-\xi)\mathrm{xZ}=(m'+n-\zeta)\mathrm{Xz}$
7. $m\mathrm{xY}\ +n'\mathrm{yz}=(m+n'-\zeta')\mathrm{xZ}=(n'+m-\xi')\mathrm{Xz}$
8. $m'\mathrm{Xy}\ +n'\mathrm{yz}=(m'+n'+\eta-\zeta'-\xi)\mathrm{xZ}=(m'+n'-\eta')\mathrm{Xz}$
9. $m\mathrm{XY}\ +n\mathrm{Yz}=(m+n-\eta)\mathrm{Xz}=(m+n+\eta'-\xi-\zeta')\mathrm{xZ}$
10. $m'\mathrm{xy}\ +n\mathrm{Yz}=(n+m'-\xi')\mathrm{Xz}=(m'+n-\zeta')\mathrm{xZ}$
11. $m\mathrm{XY}\ +n'\mathrm{yZ}=(m+n'-\zeta)\mathrm{Xz}=(n'+m-\xi)\mathrm{xZ}$
12. $m'\mathrm{xy}\ +n'\mathrm{yZ}=(m'+n'+\eta-\zeta-\xi')\mathrm{Xz}=(m'+n'-\eta')\mathrm{xZ}$
13. $m\mathrm{xY}\ +n\mathrm{Yz}=(m+n-\eta)\mathrm{xz}=(m+n+\eta'-\xi'-\zeta')\mathrm{XZ}$
14. $m'\mathrm{Xy}\ +n\mathrm{Yz}=(n+m'-\xi)\mathrm{xz}=(m'+n-\zeta')\mathrm{XZ}$
15. $m\mathrm{xY}\ +n'\mathrm{yZ}=(m+n'-\zeta)\mathrm{xz}=(n'+m-\xi')\mathrm{XZ}$
16. $m'\mathrm{Xy}\ +n'\mathrm{yZ}=(m'+n'+\eta-\zeta-\xi)\mathrm{xz}=(m'+n'-\eta')\mathrm{XZ}$

The syllogisms of chapter V are all particular cases of the above list, obtained as follows :—

De Morgan laws in applications - an example

Publishers of the Time

Formal Logic by De Morgan appeared in *Lardners Cabinet Cyclopaedia*, edited by Dyonisius Lardner, an Irish scientific writer, the first Professor of Natural History and Astronomy at the University of London, where he was involved in founding of this University.

The Society for the Diffusion of Useful Knowledge (SDUK), founded in 1826, was a Whiggish (political party with Tories) London organization that published inexpensive texts intended to adapt scientific and similarly high - minded material for the rapidly expanding reading public.

It was established mainly at the instigation of Lord Brougham with the objects of publishing information to people who were unable to obtain formal teaching, or who preferred self - education.

SDUK publications were intended for the working class and the middle class, as an antidote to the more radical output of the pauper presses. It was sometimes mentioned in contemporary sources as SDUK.

This Society published *The Library of Useful Knowledge*, to which George J. Boole is referring, in particular to the article by Sir John Lubbock.

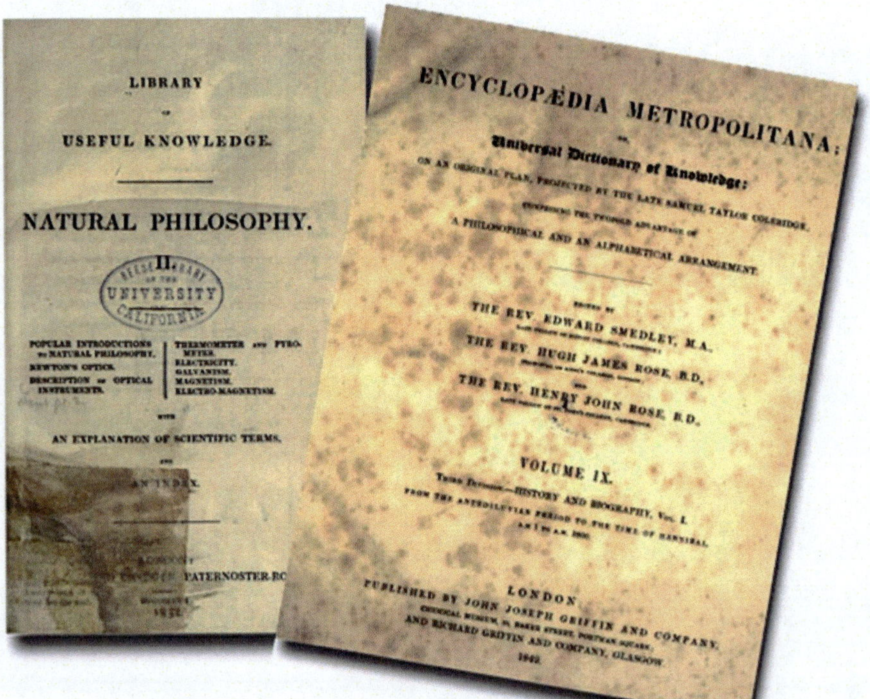

George J. Boole

2.2 Work on Logic by Boole

Boole

1847 Boole, G.J., "Mathematical Analysis of Logic, being an essay towards a calculus of deductive reasoning", London and Cambridge, spring 1847, 82 pages, Reprinted in P.E.B. Jourdain, (ed.), *George Boole's Collected Logical Works*, Vol. 1, Chicago and London, 1916.

[Boole, Ref. 1]

1848 Boole, G.J., "The calculus of logic",*The Cambridge and Dublin Mathematical Journal,* Vol. 3, 1848, 183-198, Reprinted in P.E.B.Jourdain, (ed.), *George Boole's Collected Logical Works,* Vol.1, Chicago and London, 1916.

[Boole, Ref. 2]

1854 Boole, G., J., *An Investigation of The Laws of Thought, on which are founded the mathematical theories of logic and probabilities*, 1854, v+iv+424 pages, reprinted in P.E.B.Jourdain, (ed.), *George Boole's Collected Logical Works*, Vol. 2, Chicago and London, 1916, Reprinted by Dover Publications, Inc., New York, USA, 1954.

[Boole, Ref. 3]

2.3 Mathematical Analysis of Logic

In the spring of 1847, Greorge J. Boole wrote a pamphlet entitled "Mathematical Analysis of Logic" [Boole, Ref.1], which he himself regarded as an imperfect exposition of his logical system, and always rather referred directly to the latter much elaborated treatise [Boole, Ref. 3], as a proper expression of his considerations.

That portion of this work which relates to Logic presupposes in its reader a knowledge of the most important terms of the science, as usually treated, and of its general object. On these points there is no better guide than Archbishop Whatelys Elements of Logic, *or Mr. Thomsons* Outlines of the Laws of Thought.

[Boole, Ref. 1]

That to the existing forms of Analysis a quantitative interpretation is assigned it is the result of the circumstances by which those forms were determined and is not to be construed into a universal condition of Analysis. It is upon the foundation of this general principle, that I purpose to establish the Calculus of Logic, and that I claim for it a place among the acknowledged forms of Mathematical Analysis.

[Boole, Ref. 1]

THE MATHEMATICAL ANALYSIS

OF LOGIC,

BEING AN ESSAY TOWARDS A CALCULUS
OF DEDUCTIVE REASONING.

BY GEORGE BOOLE.

Ἐπικοινωνοῦσι δὲ πᾶσαι αἱ ἐπιστῆμαι ἀλλήλαις κατὰ τὰ κοινά. Κοινὰ δὲ λέγω, οἷς χρῶνται ὡς ἐκ τούτων ἀποδεικνύντες· ἀλλ' οὐ περὶ ὧν δεικνύουσιν, οὐδὲ ὃ δεικνύουσι.

ARISTOTLE, *Anal. Post.*, lib. I. cap. XI.

CAMBRIDGE:
MACMILLAN, BARCLAY, & MACMILLAN;
LONDON: GEORGE BELL.

1847

PREFACE

In presenting this Work to public notice, I deem it not irrelevant to observe, that speculations similar to those which it records have, at different periods, occupied my thoughts. In the spring of the present year my attention was directed to the question then moved between Sir W. Hamilton and Professor De Morgan; and I was induced by the interest which it inspired, to resume the almost-forgotten thread of former inquiries. It appeared to me that, although Logic might be viewed with reference to the idea of quantity,* it had also another and a deeper system of relations. If it was lawful to regard it from *without*, as connecting itself through the medium of Number with the intuitions of Space and Time, it was lawful also to regard it from *within*, as based upon facts of another order which have their abode in the constitution of the Mind. The results of this view, and of the inquiries which it suggested, are embodied in the following Treatise.

It is not generally permitted to an Author to prescribe the mode in which his production shall be judged; but there are two conditions which I may venture to require of those who shall undertake to estimate the merits of this performance. The first is, that no preconceived notion of the impossibility of its objects shall be permitted to interfere with that candour and impartiality which the investigation of Truth demands; the second is, that their judgment of the system as a whole shall not be founded either upon the examination of only a part of it, or upon the measure of its conformity with any received system, considered as a standard of reference from which appeal is denied. It is in the general theorems which occupy the latter chapters of this work,—results to which there is no existing counterpart,—that the claims of the method, as a Calculus of Deductive Reasoning, are most fully set forth.

* See p. [42] 83.

47

A remark on the influence of De Morgan.
Comments on the independence of the results.

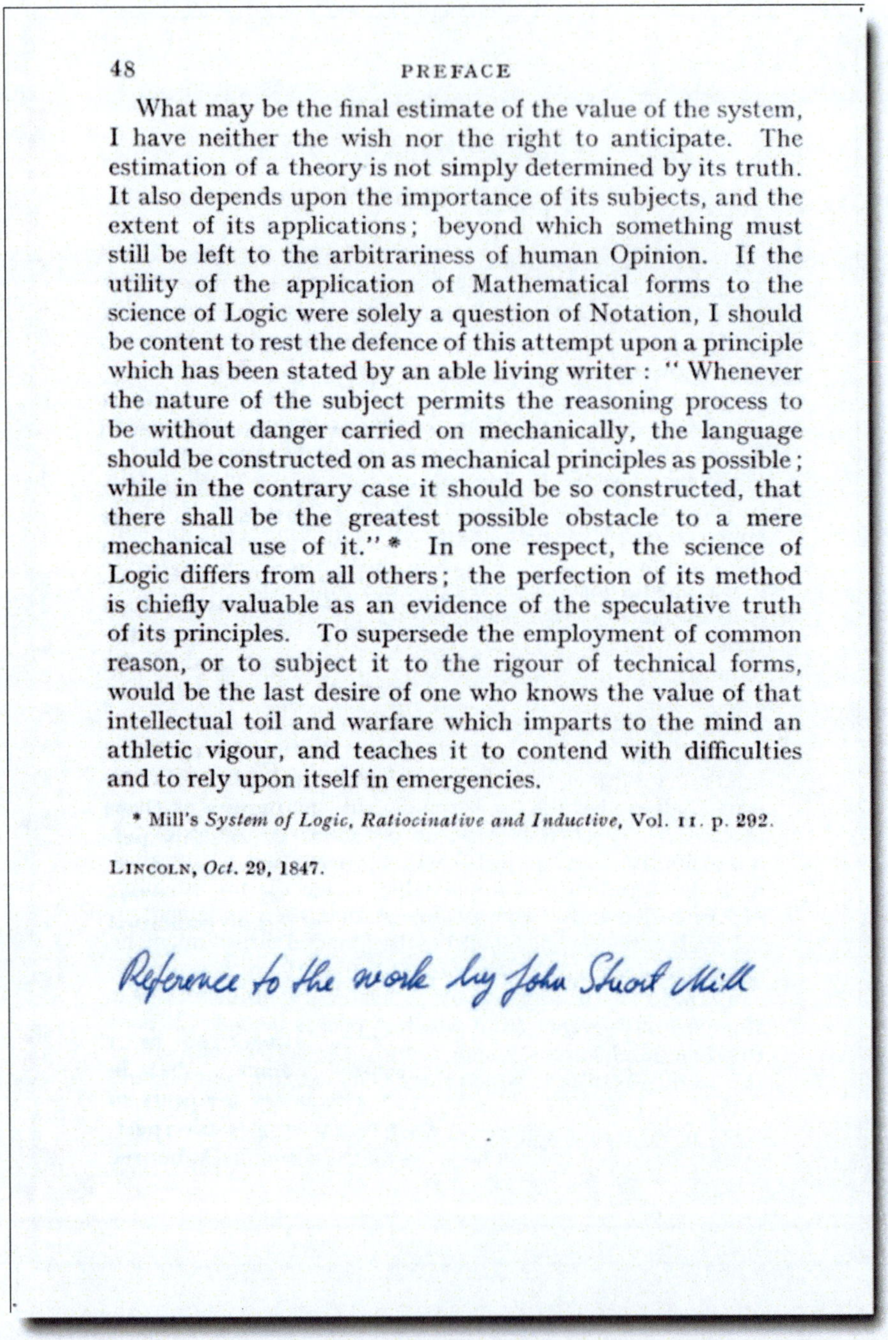

48 PREFACE

What may be the final estimate of the value of the system, I have neither the wish nor the right to anticipate. The estimation of a theory is not simply determined by its truth. It also depends upon the importance of its subjects, and the extent of its applications; beyond which something must still be left to the arbitrariness of human Opinion. If the utility of the application of Mathematical forms to the science of Logic were solely a question of Notation, I should be content to rest the defence of this attempt upon a principle which has been stated by an able living writer: "Whenever the nature of the subject permits the reasoning process to be without danger carried on mechanically, the language should be constructed on as mechanical principles as possible; while in the contrary case it should be so constructed, that there shall be the greatest possible obstacle to a mere mechanical use of it."* In one respect, the science of Logic differs from all others; the perfection of its method is chiefly valuable as an evidence of the speculative truth of its principles. To supersede the employment of common reason, or to subject it to the rigour of technical forms, would be the last desire of one who knows the value of that intellectual toil and warfare which imparts to the mind an athletic vigour, and teaches it to contend with difficulties and to rely upon itself in emergencies.

* Mill's *System of Logic, Ratiocinative and Inductive*, Vol. II. p. 292.

LINCOLN, *Oct.* 29, 1847.

Reference to the work by John Stuart Mill

Reference to the work by John Stuart Mill.

2.4 The Calculus of Logic

In [Boole, Ref. 2] in 1848, Boole wrote as the introductory statements

In a work lately published, I have exhibited the application of a new and peculiar form of mathematics to the expression of the operations of the mind in reasoning. In the present essay I design to offer such an account of a portion of this treatise as may furnish a correct view of the nature of the system developed.

View by Boole

Boole has a point of view similar to that of Gottfried Leibniz, who looked for a common formal structure.

Boole succeeded in developing a common formal structure by limiting quantitative algebra to two values.

Boolean algebra – an algebraic structure that shares essential properties of both set operations and logic operations.

The work has been done independently on other works by logicians and mathematicians at that time. For instance, results by Augustus De Morgan were not used since Boole did not consider conjunction and disjunction as a pair of dual operations.

George Boole, "The Calculus of Logic", *Cambridge and Dublin Mathematical Journal*, Vol. 3, 1848, 183-198.

II

THE CALCULUS OF LOGIC *

In a work lately published † I have exhibited the application of a new and peculiar form of Mathematics to the expression of the operations of the mind in reasoning. In the present essay I design to offer such an account of a portion of this treatise as may furnish a correct view of the nature of the system developed. I shall endeavour to state distinctly those positions in which its characteristic distinctions consist, and shall offer a more particular illustration of some features which are less prominently displayed in the original work. The part of the system to which I shall confine my observations is that which treats of categorical propositions, and the positions which, under this limitation, I design to illustrate, are the following :—

(1) That the business of Logic is with the relations of classes, and with the modes in which the mind contemplates those relations.

(2) That antecedently to our recognition of the existence of propositions, there are laws to which the conception of a class is subject—laws which are dependent upon the constitution of the intellect, and which determine the character and form of the reasoning process.

(3) That those laws are capable of mathematical expression, and that they thus constitute the basis of an interpretable calculus.

(4) That those laws are, furthermore, such that all equations which are formed in subjection to them, even though expressed under functional signs, admit of perfect solution, so that every problem in logic can be solved by reference to a general theorem.

(5) That the forms under which propositions are actually

* [*The Cambridge & Dublin Mathematical Journal*, May 1848.]

† *The Mathematical Analysis of Logic, being an Essay towards a Calculus of Deductive Reasoning.* Cambridge, Macmillan : London, G. Bell.

125

Remark on the improvements of discussions in *The Mathematical Analysis of Logic*. Important for understanding of the attitude of Boole towards his work.

2.5 The Laws of Thought

The following work is not a republication of
a former treatise by the Author, entitled
The Mathematical Analysis of Logic.
Its earlier portion is indeed devoted to the
same object, and it begins by establishing the same system of
fundamental laws, but its methods are
more general, and its range of applications
far wider. It exhibits the results,
matured by some years of study and
reflection, of a principle of investigation relating
to the intellectual operations,
the previous exposition of which was
written within a few weeks after its idea
had been conceived.

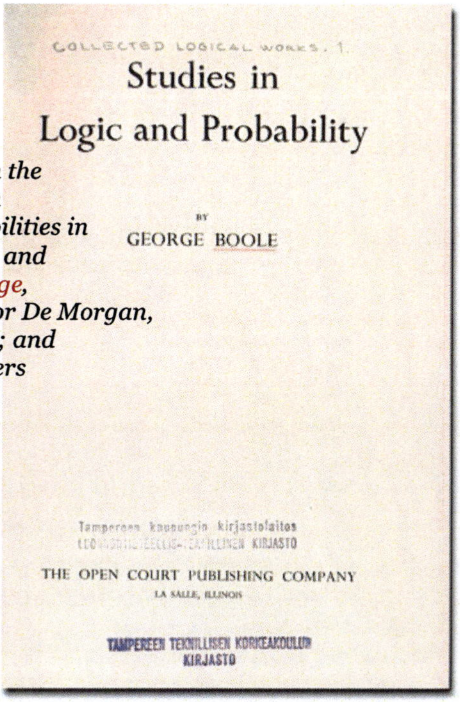
COLLECTED LOGICAL WORKS. 1.

Studies in Logic and Probability

BY

GEORGE BOOLE

Tampereen kaupungin kirjastolaitos

THE OPEN COURT PUBLISHING COMPANY

LA SALLE, ILLINOIS

TAMPEREEN TEKNILLISEN KORKEAKOULUN KIRJASTO

Preliminary information upon the
subject-matter will be found in
the special treatises on Probabilities in
Lardner's Cabinet Cyclopædia, and
the Library of Useful Knowledge,
the former of these by Professor De Morgan,
the latter by Sir John Lubbock; and
in an interesting series of Letters
translated from
the French of M. Quetelet.

[Boole, Ref. 3]

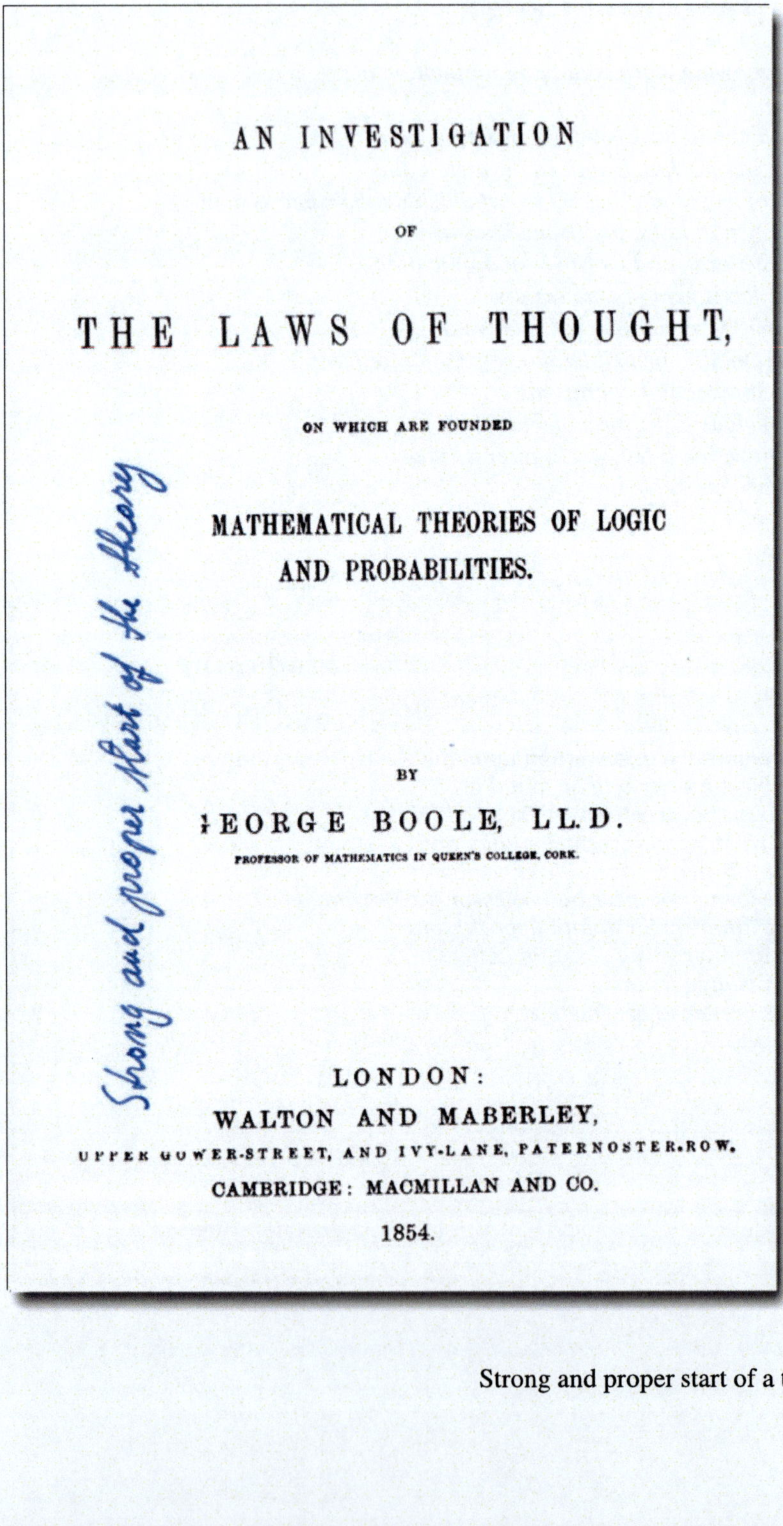

AN INVESTIGATION

OF

THE LAWS OF THOUGHT,

ON WHICH ARE FOUNDED

MATHEMATICAL THEORIES OF LOGIC
AND PROBABILITIES.

BY

GEORGE BOOLE, LL.D.

PROFESSOR OF MATHEMATICS IN QUEEN'S COLLEGE, CORK.

LONDON:
WALTON AND MABERLEY,
UPPER GOWER-STREET, AND IVY-LANE, PATERNOSTER-ROW.
CAMBRIDGE: MACMILLAN AND CO.
1854.

Strong and proper start of a theory.

Reference to the work by Whately and Thomson, and importance of publications as the Lardner's Cabinet Cyclopaedia and the Library of Useful Knowledge.

[iii]

PREFACE

THE following work is not a republication of a former treatise by the Author, entitled *The Mathematical Analysis of Logic.* Its earlier portion is indeed devoted to the same object, and it begins by establishing the same system of fundamental laws, but its methods are more general, and its range of applications far wider. It exhibits the results, matured by some years of study and reflection, of a principle of investigation relating to the intellectual operations, the previous exposition of which was written within a few weeks after its idea had been conceived.

That portion of this work which relates to Logic presupposes in its reader a knowledge of the most important terms of the science, as usually treated, and of its general object. On these points there is no better guide than Archbishop Whately's *Elements of Logic*, or Mr. Thomson's *Outlines of the Laws of Thought.* To the former of these treatises, the present revival of attention to this class of studies seems in a great measure due. Some acquaintance with the principles of Algebra is also requisite, but it is not necessary that this application should have been carried beyond the solution of simple equations. For the study of those chapters which relate to the theory of probabilities, a somewhat larger knowledge of Algebra is required, and espe-[iv]cially of the doctrine of Elimination, and of the solution of Equations containing more than one unknown quantity. Preliminary information upon the subject-matter will be found in the special treatises on Probabilities in *Lardner's Cabinet Cyclopædia*, and the *Library of Useful*

xi

Reference to the work by Whately and Thompson, and importance of publications as the Lardner's Cabinet Cyclopaedia and the Library of Useful Knowledge

End of the Preface of *The Laws of Thought*, a remark how to accept the considerations in it.

PREFACE xiii

this conflict of opinions the same truth will but emerge the more free from any personal admixture; that its different parts will be seen in their just proportion; and that none of them will eventually be too highly valued or too lightly regarded because of the prejudices which may attach to the mere form of its exposition.

To his valued friend, the Rev. George Stephens Dickson, of Lincoln, the Author desires to record his obligations for much kind assistance in the revision of this work, and for some important suggestions.

5, GRENVILLE-PLACE, CORK,
Nov. 30th, 1853.

End of the Preface of The Laws of Thought
a remark on how to accept the considerations
in it.

Dedication and Acknowledgments

To his valued friend, the Rev. George Stephens Dickson, of Lincoln, the Author desires to record his obligations for much kind assistance in the revision of this work, and for some important suggestions.

5, Grenville - place, Cork,
Nov. 30th. 1853.

TO

JOHN RYALL, LL.D.

VICE-PRESIDENT AND PROFESSOR OF GREEK

IN QUEEN'S COLLEGE, CORK

THIS WORK IS INSCRIBED

IN TESTIMONY OF FRIENDSHIP AND ESTEEM

Boole did not view logic strictly as a branch of mathematics, but emphasized the analogy between the symbols of algebra and symbols that can be used to represent logical forms and syllogisms. By unity Boole denoted the universe of thinkable objects, while literal symbols were used to attach elective meanings to common adjectives and substantives. With the use of such symbols, deriving syllogistic conclusion can be expressed in form of equations.

Manipulation with logic expressions should be used to

1. Demonstrate the truth value of a statement,
2. Rephrase a complicated statement in a simpler, more convenient, form without changing its meaning.

The latter feature is actually the foundation for engineering applications.

In [Boole, Ref. 3], Boole wrote

The design of the following treatise is to investigate the fundamental laws of those operations of the mind by which reasoning is performed; to give expression to them in the symbolical language of a Calculus, and upon this foundation to establish the science of Logic and construct its method; and to make that method itself the basis of a general method for the application of the mathematical doctrine of Probabilities; and, finally, to collect from the various elements of truth brought to view in the course of these inquiries some probable intimations concerning the nature and constitution of the human mind.

Important remarks - Language is the instrument of reasoning
Literals and symbols of algebraic operations, as well as "=" may be used to express the operation of the language, indirectly the way of thought.

such like questions is, that in the processes of reasoning, signs stand in the place and fulfil the office of the conceptions and operations of the mind ; but that as those conceptions and operations represent things, and the connexions and relations of things, so signs represent things with their connexions and relations ; and lastly, that as signs stand in the place of the conceptions and operations of the mind, they are subject to the laws of those conceptions and operations. This view will be more fully elucidated in the next chapter ; but it here serves to explain the third of those particulars involved in the definition of a sign, viz., its subjection to fixed laws of combination depending upon the nature of its interpretation.

4. The analysis and classification of those signs by which the **[27]** operations of reasoning are conducted will be considered in the following Proposition :

PROPOSITION I

All the operations of Language, as an instrument of reasoning, may be conducted by a system of signs composed of the following elements, viz. :

1st. *Literal symbols, as x, y, &c., representing things as subjects of our conceptions.*

2nd. *Signs of operation, as +, −, ×, standing for those operations of the mind by which the conceptions of things are combined or resolved so as to form new conceptions involving the same elements.*

3rd. *The sign of identity, =.*

And these symbols of Logic are in their use subject to definite laws, partly agreeing with and partly differing from the laws of the corresponding symbols in the science of Algebra.

Let it be assumed as a criterion of the true elements of rational discourse, that they should be susceptible of combination in the simplest forms and by the simplest laws, and thus combining should generate all other known and

An expression of the Laws of Thought.

32 *THE LAWS OF THOUGHT*

In the case of x representing white things, and y sheep, either of the members of this equation will represent the class of "white sheep." There may be a difference as to the order in which the conception is formed, but there is none as to the individual things which are comprehended under it. In like manner, if x represent "estuaries," and y "rivers," the expressions xy and yx will indifferently represent "rivers that are estuaries," or "estuaries that are rivers," the combination in this case being in ordinary language that of two substantives, instead of that of a substantive and an adjective as in the previous instance. Let there be a third symbol, as z, representing that class of things to which the term "navigable" is applicable, and any one of the following expressions,

$$zxy, \quad zyx, \quad xyz, \quad \&c.,$$

will represent the class of "navigable rivers that are estuaries."

If one of the descriptive terms should have some implied reference to another, it is only necessary to include that reference expressly in its stated meaning, in order to render the above remarks still applicable. Thus, if x represent "wise" and y "counsellor," we shall have to define whether x implies wisdom in the absolute sense, or only the wisdom of counsel. With such definition the law $xy = yx$ continues to be valid.

We are permitted, therefore, to employ the symbols x, y, z, &c., in the place of the substantives, adjectives, and descriptive phrases subject to the rule of interpretation, that any expression in which several of these symbols are written together shall represent all the objects or indi-[30]*viduals to which their several meanings are together applicable, and to the law that the order in which the symbols succeed each other is indifferent.*

As the rule of interpretation has been sufficiently exemplified, I shall deem it unnecessary always to express the subject "things" in defining the interpretation of a symbol used for an adjective. When I say, let x represent

A difference between the law of thought and the law of things.
The law $x \cdot y = y \cdot x$ that is used latter.

"good," it will be understood that x only represents "good" when a subject for that quality is supplied by another symbol, and that, used alone, its interpretation will be "good things."

8. Concerning the law above determined, the following observations, which will also be more or less appropriate to certain other laws to be deduced hereafter, may be added.

First, I would remark, that this law is a law of thought, and not, properly speaking, a law of things. Difference in the order of the qualities or attributes of an object, apart from all questions of causation, is a difference in conception merely. The law (1) expresses as a general truth, that the same thing may be conceived in different ways, and states the nature of that difference; and it does no more than this.

Secondly, As a law of thought, it is actually developed in a law of Language, the product and the instrument of thought. Though the tendency of prose writing is toward uniformity, yet even there the order of sequence of adjectives, absolute in their meaning, and applied to the same subject, is indifferent, but poetic diction borrows much of its rich diversity from the extension of the same lawful freedom to the substantive also. The language of Milton is peculiarly distinguished by this species of variety. Not only does the substantive often precede the adjectives by which it is qualified, but it is frequently placed in their midst. In the first few lines of the invocation to Light, we meet with such examples as the following:

> "*Offspring of heaven first-born.*"
> "The rising world of *waters dark and deep.*"
> "Bright effluence of *bright essence increate.*"

Now these inverted forms are not simply the fruits of a poetic license. They are the natural expressions of a freedom sanc-**[31]**tioned by the intimate laws of thought, but for reasons of convenience not exercised in the ordinary use of language.

Thirdly, The law expressed by (1) may be characterized

Very basic statements of the Boolean algebra, $x \cdot y = y \cdot x$, and $x \cdot x = x$.

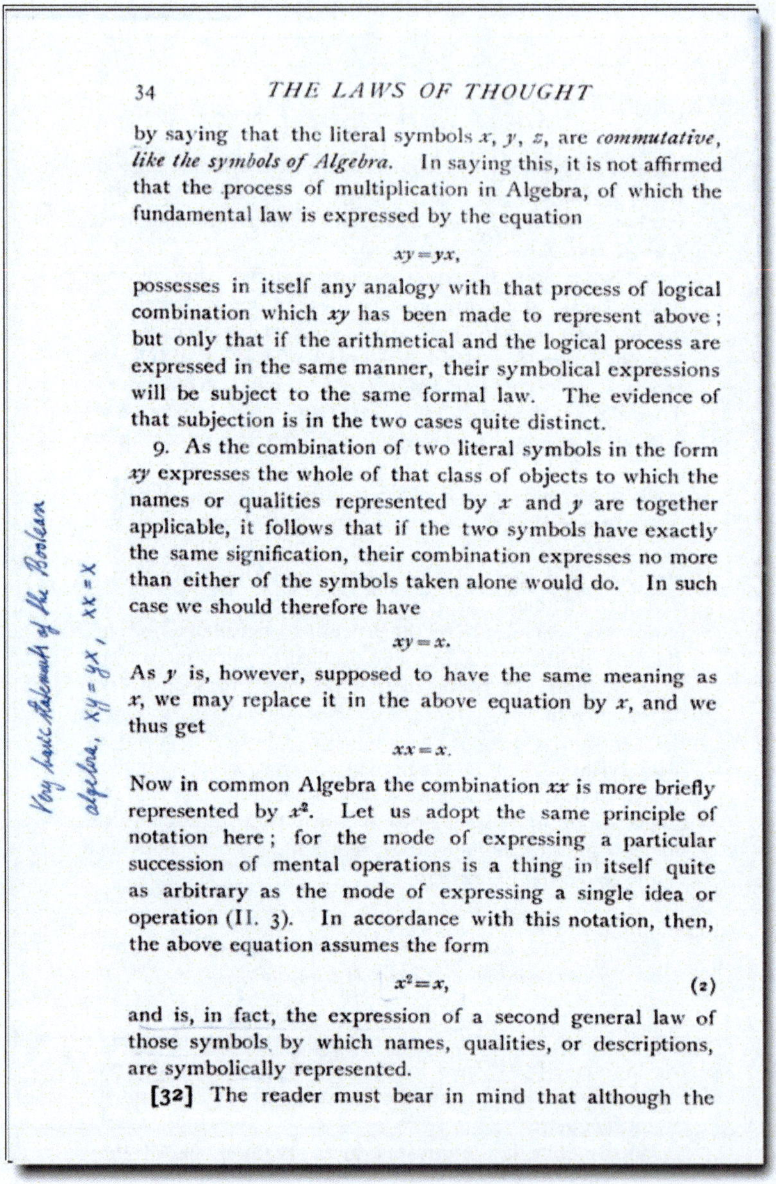

34 *THE LAWS OF THOUGHT*

by saying that the literal symbols x, y, z, are *commutative, like the symbols of Algebra.* In saying this, it is not affirmed that the process of multiplication in Algebra, of which the fundamental law is expressed by the equation

$$xy = yx,$$

possesses in itself any analogy with that process of logical combination which xy has been made to represent above; but only that if the arithmetical and the logical process are expressed in the same manner, their symbolical expressions will be subject to the same formal law. The evidence of that subjection is in the two cases quite distinct.

9. As the combination of two literal symbols in the form xy expresses the whole of that class of objects to which the names or qualities represented by x and y are together applicable, it follows that if the two symbols have exactly the same signification, their combination expresses no more than either of the symbols taken alone would do. In such case we should therefore have

$$xy = x.$$

As y is, however, supposed to have the same meaning as x, we may replace it in the above equation by x, and we thus get

$$xx = x.$$

Now in common Algebra the combination xx is more briefly represented by x^2. Let us adopt the same principle of notation here; for the mode of expressing a particular succession of mental operations is a thing in itself quite as arbitrary as the mode of expressing a single idea or operation (II. 3). In accordance with this notation, then, the above equation assumes the form

$$x^2 = x, \qquad (2)$$

and is, in fact, the expression of a second general law of those symbols by which names, qualities, or descriptions, are symbolically represented.

[32] The reader must bear in mind that although the

At this page, Boole used the 0 to denote false and 1 for true.

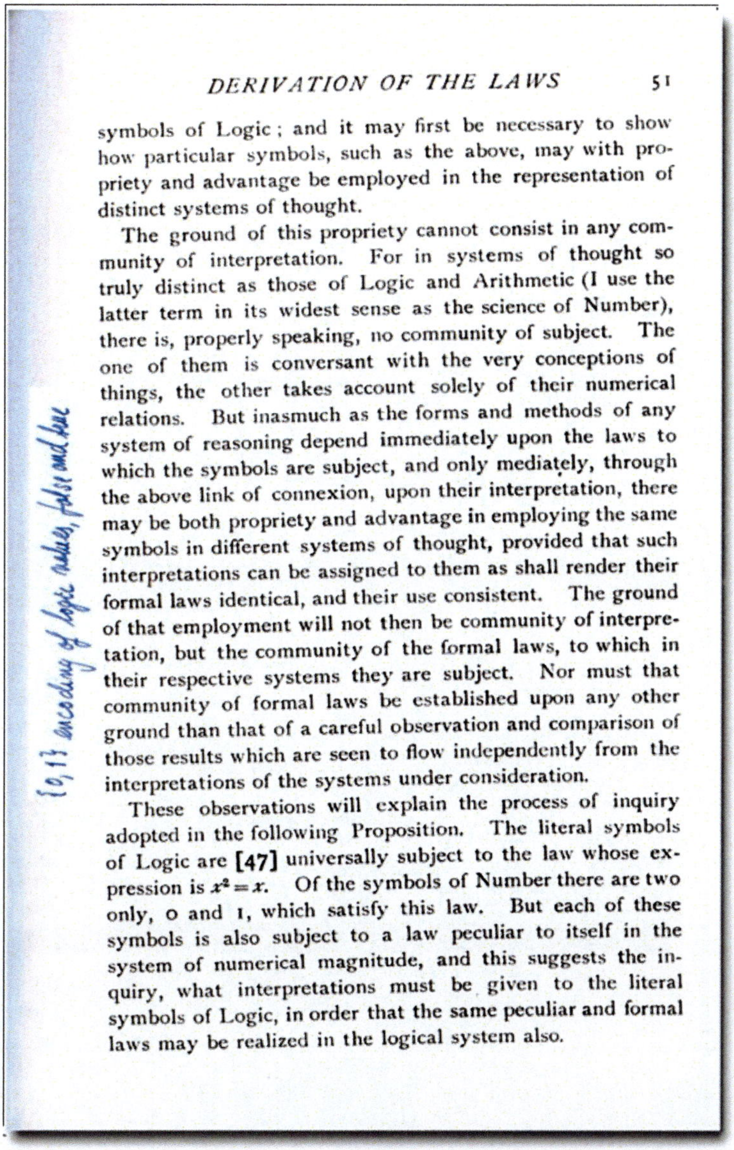

DERIVATION OF THE LAWS 51

symbols of Logic; and it may first be necessary to show how particular symbols, such as the above, may with propriety and advantage be employed in the representation of distinct systems of thought.

The ground of this propriety cannot consist in any community of interpretation. For in systems of thought so truly distinct as those of Logic and Arithmetic (I use the latter term in its widest sense as the science of Number), there is, properly speaking, no community of subject. The one of them is conversant with the very conceptions of things, the other takes account solely of their numerical relations. But inasmuch as the forms and methods of any system of reasoning depend immediately upon the laws to which the symbols are subject, and only mediately, through the above link of connexion, upon their interpretation, there may be both propriety and advantage in employing the same symbols in different systems of thought, provided that such interpretations can be assigned to them as shall render their formal laws identical, and their use consistent. The ground of that employment will not then be community of interpretation, but the community of the formal laws, to which in their respective systems they are subject. Nor must that community of formal laws be established upon any other ground than that of a careful observation and comparison of those results which are seen to flow independently from the interpretations of the systems under consideration.

These observations will explain the process of inquiry adopted in the following Proposition. The literal symbols of Logic are [47] universally subject to the law whose expression is $x^2 = x$. Of the symbols of Number there are two only, 0 and 1, which satisfy this law. But each of these symbols is also subject to a law peculiar to itself in the system of numerical magnitude, and this suggests the inquiry, what interpretations must be given to the literal symbols of Logic, in order that the same peculiar and formal laws may be realized in the logical system also.

This is how we got 0,1 encoding for logic values?

2.6 Boolean Algebra

Definition 1 *Consider a set B of at least two distinct elements* 0 *and* 1. *Assume that there are defined two binary operations* $\wedge$ *and* $\vee$, *and the unary operation* $-$ *on B, usually called logic disjunction (OR), conjunction (AND) and negation (NOT).*

An algebraic system $\langle B, \wedge, \vee, -, 0, 1\rangle$ *is a Boolean algebra if for any the following axioms are satisfied*

Idempotence	$a \vee a = a, \quad a \cdot a = a$
Comutativity	$a \vee b = b \vee a, \quad a \cdot b = b \cdot a$
Associativity	$a \vee (b \vee c) = (a \vee b) \vee c,$ $a \cdot (b \cdot c) = (a \cdot b) \cdot c$
Absorption	$a \vee (a \cdot b) = a, \quad a \cdot (a \vee b) = a$
Distributivity	$a \vee (b \cdot c) = (a \vee b) \cdot (a \vee c)$ $a \cdot (b \vee c) = (a \cdot b) \vee (a \cdot c)$
Involutivity	$\overline{\overline{a}} = a$
Complement	$a \vee \overline{a} = 1, \quad a \cdot \overline{a} = 0$ $a \vee 0 = a, \quad a \cdot 1 = a$
Identity	$a \vee 1 = 1, \quad a \cdot 0 = 0$
De Morgan Laws	$\overline{(a \vee b)} = \overline{a} \cdot \overline{b}, \quad \overline{a \cdot b} = \overline{a} \vee \overline{b}$

William Stanley Jevons

2.7 Work on Logic by Jevons

Jevons first work on logic in 1864 sold four samples in six months.

Jevons performed a function specified by a truth table by using his Logic Machine, see below. He was the first user of matrix analysis as stated by

W. Mays and D. P. Henry, "Jevons and Logic", *Mind, A Quarterly Review of Psychology and Philosophy*, T. Nelson & Sons, Edinburgh, Vol. LXII, 1953, 448-505.

Jevons

1864 *Pure Logic or the Logic of Quality apart from Quantity with Remarks on Boole's System and the Relation of Logic and Mathematics*, E. Stanford, London
Substitution of Similars, Macmillan, London

1869 "On the mechanical performance of logical inference", *Philosophical Transactions of the Royal Society*, Vol.
1870 160, 497-518.

The Principles of Science, A Treatise on Logic and Scientific Method 2 vols., Mcmillan & Co., London, 2nd
1874 ed., Mcmillan & Co., London and New York

Studies in Deductive Logic - A Manual for Students, Mcmillan & Co., London

1880

Pure Logic and Other Minor Works, Pure Logic or the Logic of Quality Apart From the Quantity,
1890 Macmillan and Co., London

Pure Logic and Other Minor Works, Robert Adamson and Harriet A. Jevons, (eds.), Mcmillan & Co., London and New York, 1890, reprinted by Thoemmes Press, Bristol, 1991.

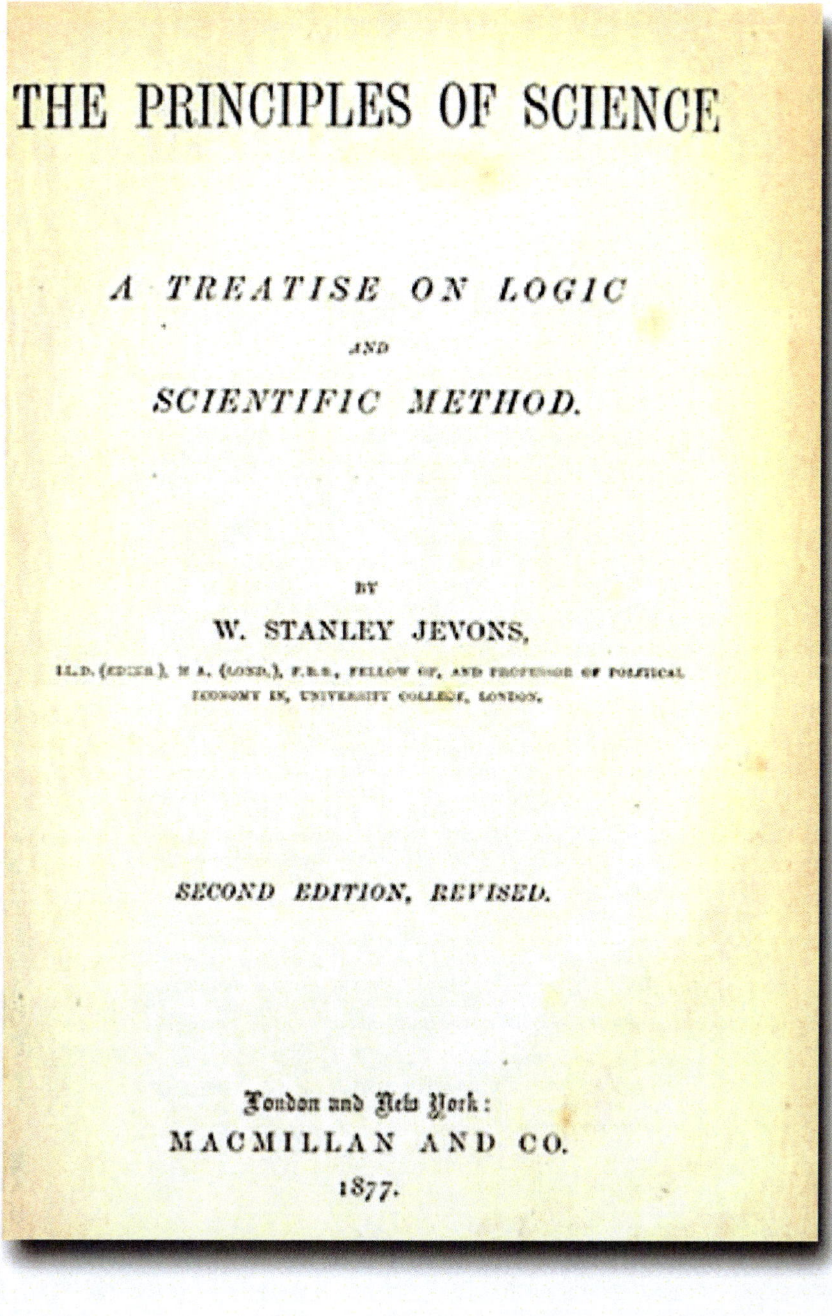

THE PRINCIPLES OF SCIENCE

A TREATISE ON LOGIC

AND

SCIENTIFIC METHOD.

BY

W. STANLEY JEVONS,

LL.D. (EDINB.), M.A. (LOND.), F.R.S., FELLOW OF, AND PROFESSOR OF POLITICAL ECONOMY IN, UNIVERSITY COLLEGE, LONDON.

SECOND EDITION, REVISED.

London and New York:
MACMILLAN AND CO.
1877.

Attempts towards disclosing a general formula of logic inference.

sameness. For this purpose we may generalise in like manner the symbol $\sim$, which was introduced by Wallis to signify difference between quantities. The general formula

$$B \sim C$$

denotes that B and C are the names of two objects or groups which are not identical with each other. Thus we may say

Acrogens $\sim$ Flowering plants.
Snowdon $\sim$ The highest mountain in Great Britain.

I shall also occasionally use the sign $\backsim$ to signify in the most general manner the existence of any relation between the two terms connected by it. Thus $\backsim$ might mean not only the relations of equality or inequality, sameness or difference, but any special relation of time, place, size, causation, &c. in which one thing may stand to another. By $A \backsim B$ I mean, then, any two objects of thought related to each other in any conceivable manner.

General Formula of Logical Inference.

The one supreme rule of inference consists, as I have said, in the direction to affirm of anything whatever is known of its like, equal or equivalent. The *Substitution of Similars* is a phrase which seems aptly to express the capacity of mutual replacement existing in any two objects which are like or equivalent to a sufficient degree. It is matter for further investigation to ascertain when and for what purposes a degree of similarity less than complete identity is sufficient to warrant substitution. For the present we think only of the exact sameness expressed in the form

$$A = B.$$

Now if we take the letter C to denote any third conceivable object, and use the sign $\backsim$ in its stated meaning of *indefinite relation*, then the general formula of all inference may be thus exhibited:—

From $A = B \backsim C$
we may infer $A \backsim C$

or, in words—*In whatever relation a thing stands to a second thing, in the same relation it stands to the like or equivalent of that second thing.* The identity between A

Importance of inference, conclusions from partial identities.

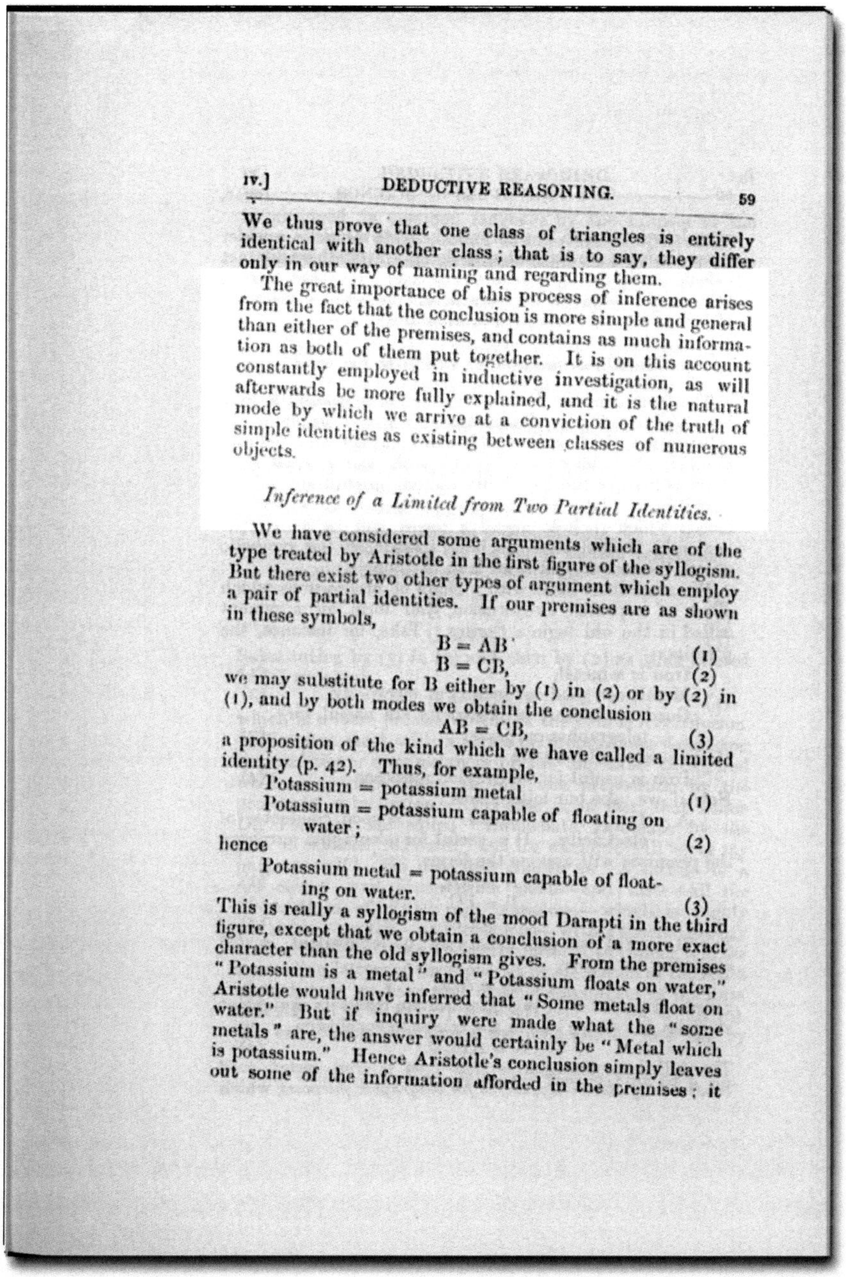

We thus prove that one class of triangles is entirely identical with another class; that is to say, they differ only in our way of naming and regarding them.

The great importance of this process of inference arises from the fact that the conclusion is more simple and general than either of the premises, and contains as much information as both of them put together. It is on this account constantly employed in inductive investigation, as will afterwards be more fully explained, and it is the natural mode by which we arrive at a conviction of the truth of simple identities as existing between classes of numerous objects.

Inference of a Limited from Two Partial Identities.

We have considered some arguments which are of the type treated by Aristotle in the first figure of the syllogism. But there exist two other types of argument which employ a pair of partial identities. If our premises are as shown in these symbols,

$$B = AB \quad (1)$$

$$B = CB, \quad (2)$$

we may substitute for B either by (1) in (2) or by (2) in (1), and by both modes we obtain the conclusion

$$AB = CB, \quad (3)$$

a proposition of the kind which we have called a limited identity (p. 42). Thus, for example,

Potassium = potassium metal (1)

Potassium = potassium capable of floating on water; (2)

hence

Potassium metal = potassium capable of floating on water. (3)

This is really a syllogism of the mood Darapti in the third figure, except that we obtain a conclusion of a more exact character than the old syllogism gives. From the premises "Potassium is a metal" and "Potassium floats on water," Aristotle would have inferred that "Some metals float on water." But if inquiry were made what the "some metals" are, the answer would certainly be "Metal which is potassium." Hence Aristotle's conclusion simply leaves out some of the information afforded in the premises; it

Numerically definite reasoning, again Mathematics to Logic.

hardly requisite to decide how we define the names of numbers, provided we remember that out of the infinitely numerous relations of one number to others, some one relation expressed in an equality must be a definition of the number in question and the other relations immediately become necessary inferences.

In the science of number the variety of classes which can be formed is altogether infinite, and statements of perfect generality may be made subject only to difficulty or exception at the lower end of the scale. Every existing number for instance belongs to the class $m + 7$; that is, every number must be the sum of another number and seven, except of course the first six or seven numbers, negative quantities not being here taken into account. Every number is the half of some other, and so on. The subject of generalization, as exhibited in mathematical truths, is an infinitely wide one. In number we are only at the first step of an extensive series of generalizations. As number is general compared with the particular things numbered, so we have general symbols for numbers, and general symbols for relations between undetermined numbers. There is an unlimited hierarchy of successive generalizations.

Numerically Definite Reasoning.

It was first discovered by De Morgan that many arguments are valid which combine logical and numerical reasoning, although they cannot be included in the ancient logical formulas. He developed the doctrine of the "Numerically Definite Syllogism," fully explained in his *Formal Logic* (pp. 141—170). Boole also devoted considerable attention to the determination of what he called "Statistical Conditions," meaning the numerical conditions of logical classes. In a paper published among the Memoirs of the Manchester Literary and Philosophical Society, Third Series, vol. IV. p. 330 (Session 1869—70), I have pointed out that we can apply arithmetical calculation to the Logical Alphabet. Having given certain logical conditions and the numbers of objects in certain classes, we can either determine the numbers of objects in other classes governed by those conditions, or can show what

2.8 Jevons and Boole

Jevons has been strongly influenced by Boole, as he explicitly stated in the second edition of *The Principles of Science* in 1877.

As to my own view of Logic, they were originally mounted by a careful study of Boole's work, as fully stated in my first logical essay

(*Pure Logic*, in 1864).

Jevons wrote

... the mathematical dress into which [Boole] threw his discoveries is not proper to them, and his quasi-mathematical processes are vastly more complicated than they need have been.

A review of the paper by George Bruce Halsted "Prof. Jevons's Criticism of Boole's Logical System" *Mind*, Vol. 3, No. 9, January 1878, 134-137
This article consists of 4 pages.

134 *Notes and Discussions.*

explicable in the same way as the interchanges above referred to. But in other classes of society real or apparent varieties of such phonetic interchange, which I have elsewhere designated "Cross Compensation" (*Grimm's Law*, Trübner, 1876) have established themselves as dialectic characteristics. Thus the plant-seller that haunts our ways all the summer vociferates "Roots for your gard*ing* all agrow*in* and ablow*in*"; and the lavender girl that follows him sings "sweet-smell*in* lav*ing*der," &c. This class of instances, therefore, offers for investigation not only an origin but a history.*

The object of this note, however, is not (as is obvious enough) to investigate these curious phenomena, so much as to suggest that they deserve investigation. If Mr. Verdon, or some other professed psychologist, would subject them to a thorough discussion, he would, besides amusing himself, instruct inquirers in other lines of study (language, for example) which, without being purely psychological, necessitate a frequent reference to psychological principles.

T. LE M. DOUSE.

Prof. Jevons's criticism of Boole's Logical System.—The appearance of a new edition of Prof. Jevons's *Principles of Science* shows that his partial adaptation of Boole's system has gained a wider circulation than its original, and renders not inopportune a few words on the two men.

In the preface to this second edition Prof. Jevons says: "As to my own views of Logic, they were originally moulded by a careful study of Boole's works, as fully stated in my first logical essay". So it has seemed best to me to go back to this *Pure Logic* of 1864, and taking his first and last works together, to discuss carefully his criticisms of Boole. In both books one is struck by the fact that Prof. Jevons has never risen from the conception of the old Algebra of Number to the idea of Algebras in general. For him "all the wondrous branches of mathematical calculus" are merely developed Arithmetic (*P. of S.*, p. 162). Yet he appreciates the importance of Descartes' mathematical discovery without noting that it was really making a new Algebra, the Algebra of Geometry, introducing the directed line, the variable, &c., and not being a mere outgrowth from the old Algebra of Number. He mentions also the new Algebra of Quaternions, which contains laws flatly contradicting those of number, yet he does not draw the obvious conclusion. Finally, though Boole's Algebra of Logic is founded on the condition $x^2 = x$ or $x(1 - x) = 0$, which is not true of numbers in general, Prof. Jevons persists in considering it "a numerical system".

What would he say of Grassmann's system, of Mr. Spottiswoode's

* Many examples may be collected by the student of English popular idioms. A collection from the German dialects has recently appeared in Herr J. F. Kräuter's treatise *Die Lautverschiebung*, pp. 60-62.

A Review by Carveth Read of *Pure Logic and Other Minor Works* by W. Stanley Jevons, Robert Adamson, Harriett A. Jevons, has been published in *Mind*, Vol. 16, No. 61, January 1891, 106-110.

2.9 Mechanization of Boolean Logic

Jevons had an idea to construct devices that will support logical thinking in the same same way as, for example, abacus helps in numerical calculations.

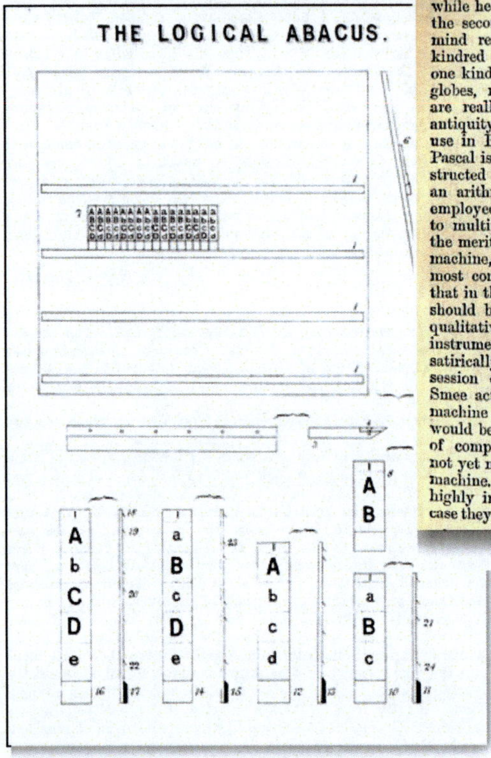

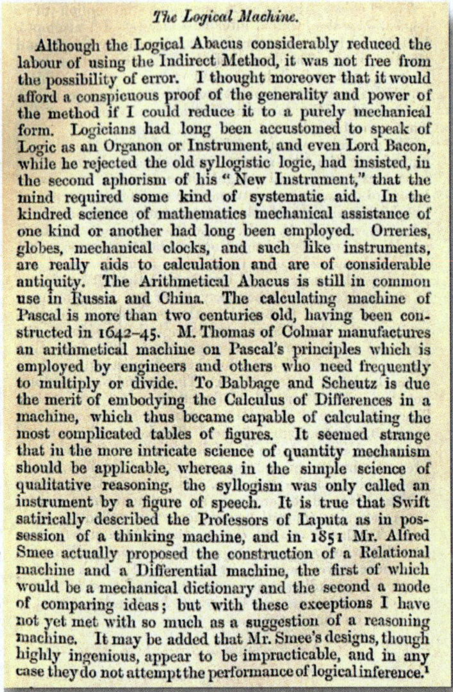

The Logical Machine.

Although the Logical Abacus considerably reduced the labour of using the Indirect Method, it was not free from the possibility of error. I thought moreover that it would afford a conspicuous proof of the generality and power of the method if I could reduce it to a purely mechanical form. Logicians had long been accustomed to speak of Logic as an Organon or Instrument, and even Lord Bacon, while he rejected the old syllogistic logic, had insisted, in the second aphorism of his "New Instrument," that the mind required some kind of systematic aid. In the kindred science of mathematics mechanical assistance of one kind or another had long been employed. Orreries, globes, mechanical clocks, and such like instruments, are really aids to calculation and are of considerable antiquity. The Arithmetical Abacus is still in common use in Russia and China. The calculating machine of Pascal is more than two centuries old, having been constructed in 1642–45. M. Thomas of Colmar manufactures an arithmetical machine on Pascal's principles which is employed by engineers and others who need frequently to multiply or divide. To Babbage and Scheutz is due the merit of embodying the Calculus of Differences in a machine, which thus became capable of calculating the most complicated tables of figures. It seemed strange that in the more intricate science of quantity mechanism should be applicable, whereas in the simple science of qualitative reasoning, the syllogism was only called an instrument by a figure of speech. It is true that Swift satirically described the Professors of Laputa as in possession of a thinking machine, and in 1851 Mr. Alfred Smee actually proposed the construction of a Relational machine and a Differential machine, the first of which would be a mechanical dictionary and the second a mode of comparing ideas; but with these exceptions I have not yet met with so much as a suggestion of a reasoning machine. It may be added that Mr. Smee's designs, though highly ingenious, appear to be impracticable, and in any case they do not attempt the performance of logical inference.[1]

A description of logic machine is given at the above page from *The Principles of Science*

Barrett, L., Connell, M.,"Jevons and the logic piano", *The Rutherford Journal, The New Zealand Journal for the History and Philosophy of Science and Technology*, Vol. 1, December 2005, www.rutherfordjournal.org

Logic piano

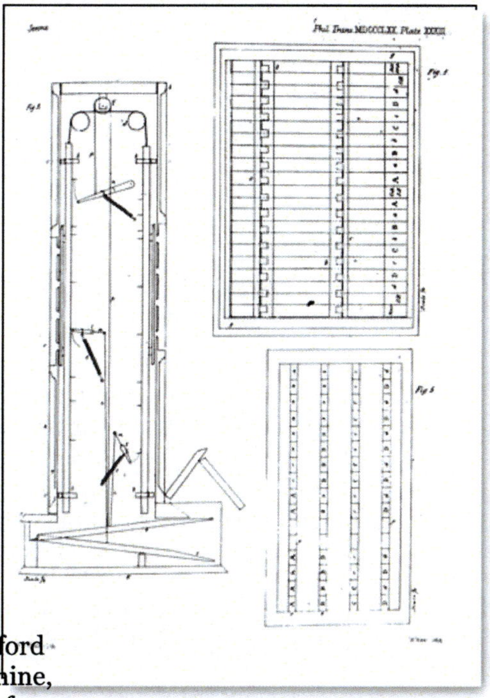

In 1869, a clockmaker from Salford made for Jevons the Logic machine, or Logic piano, which consists of a series of wooden boards with various combinations of true and false terms.

The device worked with up to 4 terms, which makes 65536 logical combinations.

The Logic piano performs operations of logic inference.

Two pages on the right are reprints from *The Principles of Science* showing details of the construction of the Logic Piano.

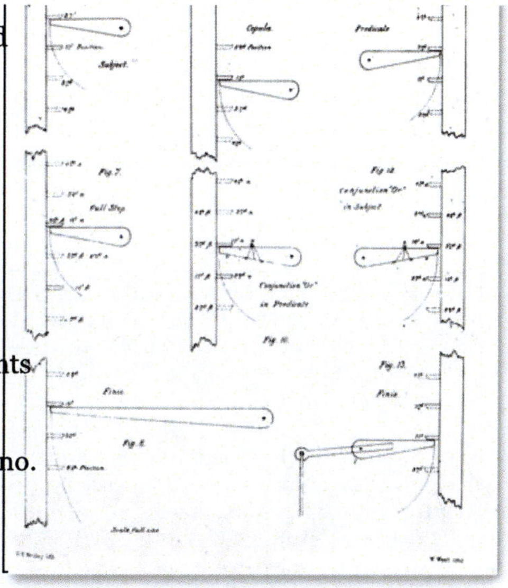

Jevons wanted to construct a logical abacus as a device capable of performing logical inference in the way similar to that as the arithmetic abacus is used for calculations, or resembling the work of calculating machine by Pascal, similar machine on the same principles set by Pascal as developed by M. Thomas of Calmar manufacturers, and *Differential Machine* by *Babbage* and *Scheutz.*

In explaining his intentions with Logic abacus, Jevons also mentioned ideas of *Alfred Smee*, who proposed a *Relational machine* as a mechanical dictionary and a Differential machine that would be able to compare ideas.

The wooden boards can be arranged on a rack and a ruler is used to remove certain excluded combinations. The movement of boards in a correct way was provided by levers and pulleys.

The logic expressions are typed (played) by pressing keys at the keyboard where pressing full stop removes all impossible combinations. The copula is used as the equal key, and finis key resets the machine.

Disjunctive propositions are to be treated in an exactly similar manner. Thus, to work the premises

$$A = AB + AC$$
$$B + C = BD + CD,$$

it is only necessary to press in succession the keys

A (left), copula, A (right), B, +, A,C, full stop.
B (left), +, C, copula, B (right), D, +, C,D, full stop.

The combinations then remaining will be as follows

ABCD	*a*BCD	*abc*D
AB*c*D	*a*B*c*D	*abcd*.
A*b*CD	*ab*CD	

The terms *A,B,C,D* and their logic complements *a,b,c,d,* were displayed at two sides of the device, at the front and the back side.

It ought to be carefully observed that the logical machine cannot receive a simple identity of the form A = B except in the double form of A = B and B = A. To work the proposition A = B, it is therefore necessary to press the keys—

A (left), copula, B (right), full stop ;
B (left), copula, A (right), full stop.

The same double operation will be necessary whenever the proposition is not of the kind called a partial identity (p. 40). Thus AB = CD, AB = AC, A = B + C, A + B = C + D, all require to be read from both ends separately.

The device he desired could work with 16 terms, however, it was large to fit in his office.

In present terminology, the Logic machine by Jevons implements functions specified by truth tables.

The device was described in *Substitution of Similars*, and also in *The Principles of Science A Treatise on Logic and Scientific Methods*, Macmillan, London, 1877.

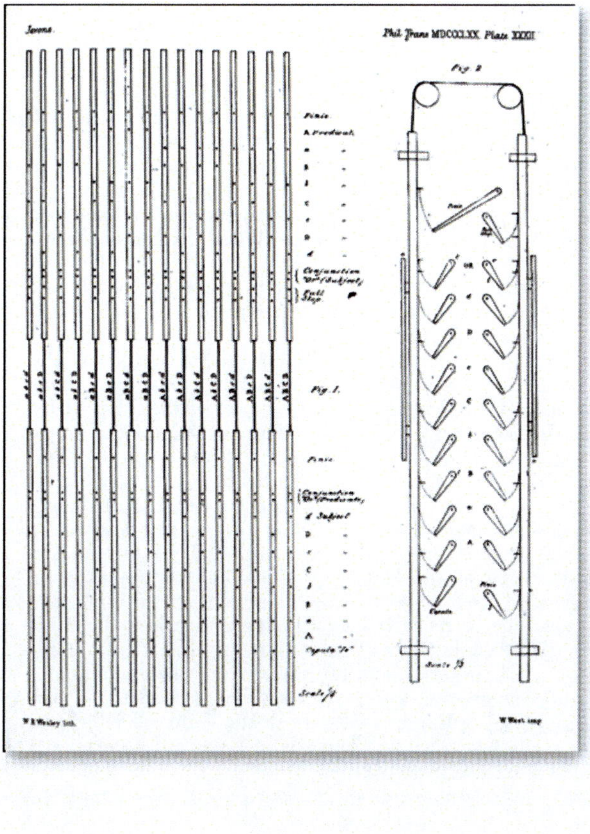

The mechanism has been protected by a wooden case three feet in height, with a keyboard to perform the substitution procedure.

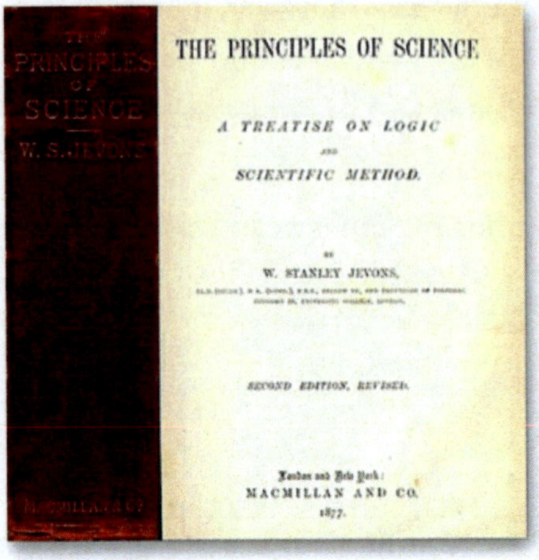

The general appearance of the machine is shown in a plate facing the title-page of this volume. It somewhat resembles a very small upright piano or organ, and has a keyboard containing twenty-one keys. These keys are of two kinds, sixteen of them representing the terms or letters A, *a*, B, *b*, C, *c*, D, *d*, which have so often been employed in our logical notation. When letters occur on the left-hand side of a proposition, formerly called the subject, each is represented by a key on the left-hand half of the keyboard; but when they occur on the right-hand side, or as it used to be called the predicate of the proposition, the letter-keys on the right-hand side of the keyboard are the proper representatives. The five other keys may be called operation keys, to distinguish them from the letter or term keys. They stand for the stops, copula, and disjunctive conjunctions of a proposition. The middle key of all is the copula, to be pressed when the verb *is* or the sign = is met. The key to the extreme right-hand is called the Full Stop, because it should be pressed when a proposition is completed, in fact in the proper place of the full stop. The key to the extreme left-hand is used to terminate an argument or to restore the machine to its initial condition; it is called the Finis key. The last keys but one on the right and left complete the whole series, and represent the conjunction *or* in its unexclusive meaning, or the sign ꖌ which I have employed, according as it occurs in the right or left hand side of the proposition. The whole keyboard is arranged as shown on the next page—

Finis.	Left-hand side of Proposition.									Copula.	Right-hand side of Proposition.									Full Stop.
	ꖌ Or	d	D	c	C	b	B	a	A		A	a	B	b	C	c	D	d	ꖌ Or	

To work the machine it is only requisite to press the keys in succession as indicated by the letters and signs of a symbolical proposition. All the premises of an argument are supposed to be reduced to the simple notation which has been employed in the previous pages. Taking then such a simple proposition as

$$A = AB,$$

we press the keys A (left), copula, A (right), B (right), and full stop.

If there be a second premise, for instance

$$B = BC,$$

we press in like manner the keys—

B (left), copula, B (right), C (right), full stop.

The process is exactly the same however numerous the premises may be. When they are completed the operator will see indicated on the face of the machine the exact combinations of letters which are consistent with the premises according to the principles of thought.

As shown in the figure opposite the title-page, the machine exhibits in front a Logical Alphabet of sixteen combinations, exactly like that of the Abacus, except that the letters of each combination are separated by a certain interval. After the above problem has been worked upon the machine the Logical Alphabet will have been modified so as to present the following appearance—

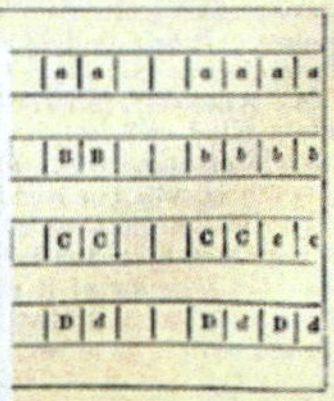

A desctiption of the Logic piano and details of its functioning written by Jevons.

A description of the logic abacus.

104 THE PRINCIPLES OF SCIENCE. [CHAP.

that is, "a fixed star is not a planet, but is either subject or not, as the case may be, to gravity." Here we have no conclusion concerning the connection of fixed stars and gravity.

The Logical Abacus.

The Indirect Method of Inference has now been sufficiently described, and a careful examination of its powers will show that it is capable of giving a full analysis and solution of every question involving only logical relations. The chief difficulty of the method consists in the great number of combinations which may have to be examined; not only may the requisite labour become formidable, but a considerable chance of mistake arises. I have therefore given much attention to modes of facilitating the work, and have succeeded in reducing the method to an almost mechanical form. It soon appeared obvious that if the conceivable combinations of the Logical Alphabet, for any number of letters, instead of being printed in fixed order on a piece of paper or slate, were marked upon light movable pieces of wood, mechanical arrangements could readily be devised for selecting any required class of the combinations. The labour of comparison and rejection might thus be immensely reduced. This idea was first carried out in the Logical Abacus, which I have found useful in the lecture-room for exhibiting the complete solution of logical problems. A minute description of the construction and use of the Abacus, together with figures of the parts, has already been given in my essay called *The Substitution of Similars,*[1] and I will here give only a general description.

The Logical Abacus consists of a common school blackboard placed in a sloping position and furnished with four horizontal and equi-distant ledges. The combinations of the letters shown in the first four columns of the Logical Alphabet are printed in somewhat large type, so that each letter is about an inch from the neighbouring one, but the letters are placed one above the other instead of being in horizontal lines as in p. 94. Each combination of letters is separately fixed to the surface of

[1] Pp. 55—59, 81—86.

More details of the logic abacus.

a thin slip of wood one inch broad and about one-eighth inch thick. Short steel pins are then driven in an inclined position into the wood. When a letter is a large capital representing a positive term, the pin is fixed in the upper part of its space; when the letter is a small italic representing a negative term, the pin is fixed in the lower part of the space. Now, if one of the series of combinations be ranged upon a ledge of the black-board, the sharp edge of a flat rule can be inserted beneath the pins belonging to any one letter—say A, so that all the combinations marked A can be lifted out and placed upon a separate ledge. Thus we have represented the act of thought which separates the class A from what is not-A. The operation can be repeated; out of the A's we can in like manner select those which are B's, obtaining the AB's; and in like manner we may select any other classes such as the *a*B's, the *ab*'s, or the *abc*'s.

If now we take the series of eight combinations of the letters A, B, C, *a*, *b*, *c*, and wish to analyse the argument anciently called Barbara, having the premises

$$A = AB \qquad (1)$$

$$B = BC, \qquad (2)$$

we proceed as follows—We raise the combinations marked *a*, leaving the A's behind; out of these A's we move to a lower ledge such as are *b*'s, and to the remaining AB's we join the *a*'s which have been raised. The result is that we have divided all the combinations into two classes, namely, the A*b*'s which are incapable of existing consistently with premise (1), and the combinations which are consistent with the premise. Turning now to the second premise, we raise out of those which agree with (1) the *b*'s, then we lower the B*c*'s; lastly we join the *b*'s to the BC's. We now find our combinations arranged as below.

A B C				*a* B C		*a* *b* C	*a* *b* *c*
	A B *c*	A *b* C	A *b* *c*		*a* B *c*		

The lower line contains all the combinations which are inconsistent with either premise; we have carried out in a

The logic machine, something more than abacus.

108 THE PRINCIPLES OF SCIENCE. [CHAP.

The Logical Abacus soon suggested the notion of a Logical Machine, which, after two unsuccessful attempts, I succeeded in constructing in a comparatively simple and effective form. The details of the Logical Machine have been fully described by the aid of plates in the Philosophical Transactions,[1] and it would be needless to repeat the account of the somewhat intricate movements of the machine in this place.

The general appearance of the machine is shown in a plate facing the title-page of this volume. It somewhat resembles a very small upright piano or organ, and has a keyboard containing twenty-one keys. These keys are of two kinds, sixteen of them representing the terms or letters A, *a*, B, *b*, C, *c*, D, *d*, which have so often been employed in our logical notation. When letters occur on the left-hand side of a proposition, formerly called the subject, each is represented by a key on the left-hand half of the keyboard; but when they occur on the right-hand side, or as it used to be called the predicate of the proposition, the letter-keys on the right-hand side of the keyboard are the proper representatives. The five other keys may be called operation keys, to distinguish them from the letter or term keys. They stand for the stops, copula, and disjunctive conjunctions of a proposition. The middle key of all is the copula, to be pressed when the verb *is* or the sign = is met. The key to the extreme right-hand is called the Full Stop, because it should be pressed when a proposition is completed, in fact in the proper place of the full stop. The key to the extreme left-hand is used to terminate an argument or to restore the machine to its initial condition; it is called the Finis key. The last keys but one on the right and left complete the whole series, and represent the conjunction *or* in its unexclusive meaning, or the sign ꖌ which I have employed, according as it occurs in the right or left hand side of the proposition. The whole keyboard is arranged as shown on the next page—

ential Machines. Also *Philosophical Transactions*, [1870] vol. 160, p. 518.

[1] *Philosophical Transactions* [1870], vol. 160, p. 497. *Proceedings of the Royal Society*, vol. xviii. p. 166, Jan. 20 1870. *Nature*, vol. i. p. 343.

An illustration of the Logic machine.

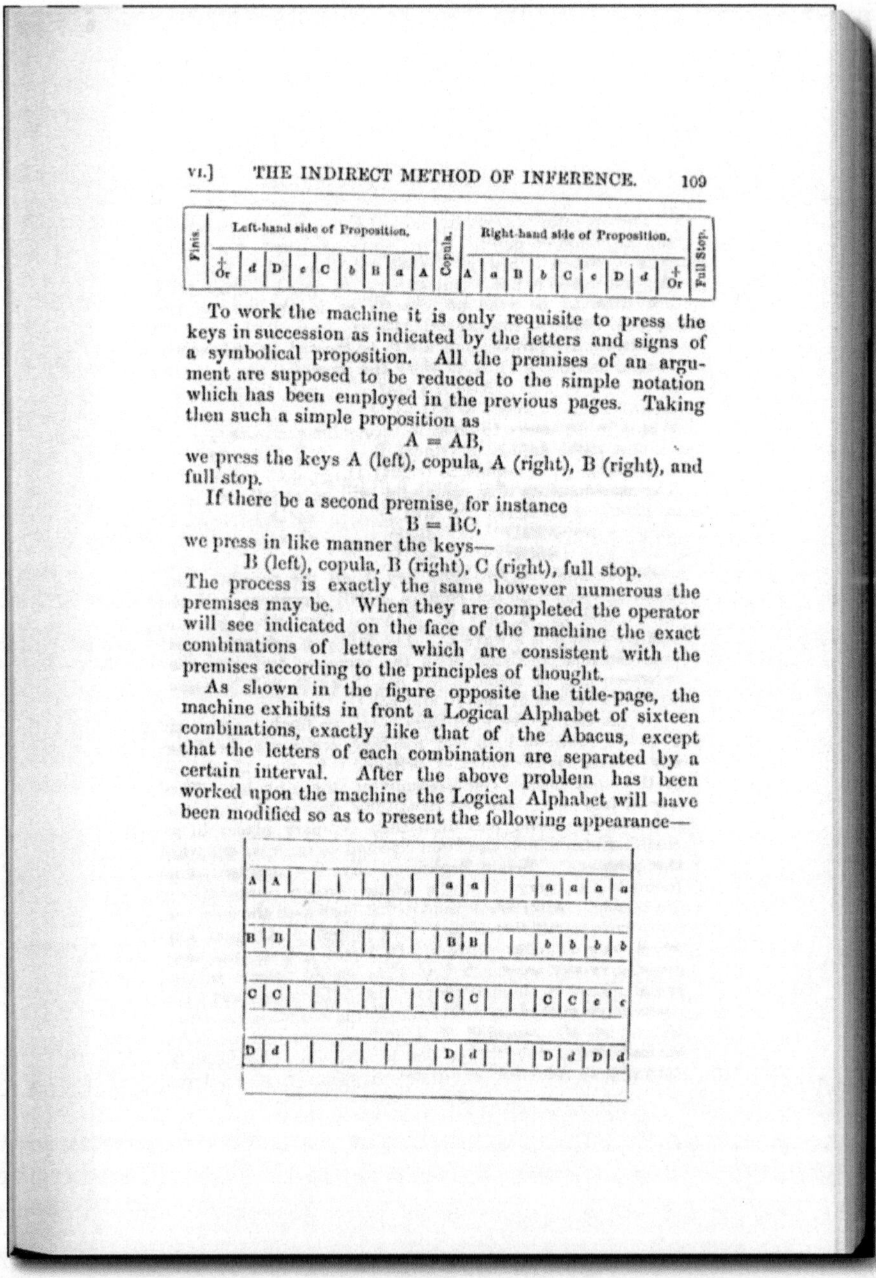

VI.] THE INDIRECT METHOD OF INFERENCE. 109

Finis.	Left-hand side of Proposition.									Copula.	Right-hand side of Proposition.									Full Stop.
	+ Or	d	D	c	C	b	B	a	A		A	a	B	b	C	c	D	d	+ Or	

To work the machine it is only requisite to press the keys in succession as indicated by the letters and signs of a symbolical proposition. All the premises of an argument are supposed to be reduced to the simple notation which has been employed in the previous pages. Taking then such a simple proposition as

A = AB,

we press the keys A (left), copula, A (right), B (right), and full stop.

If there be a second premise, for instance

B = BC,

we press in like manner the keys—

B (left), copula, B (right), C (right), full stop.

The process is exactly the same however numerous the premises may be. When they are completed the operator will see indicated on the face of the machine the exact combinations of letters which are consistent with the premises according to the principles of thought.

As shown in the figure opposite the title-page, the machine exhibits in front a Logical Alphabet of sixteen combinations, exactly like that of the Abacus, except that the letters of each combination are separated by a certain interval. After the above problem has been worked upon the machine the Logical Alphabet will have been modified so as to present the following appearance—

A	A							a	a			a	a	a	a
B	B							B	B			b	b	b	b
C	C							C	C			C	C	c	c
D	d							D	d			D	d	D	d

Reference to Jevons by Boole.

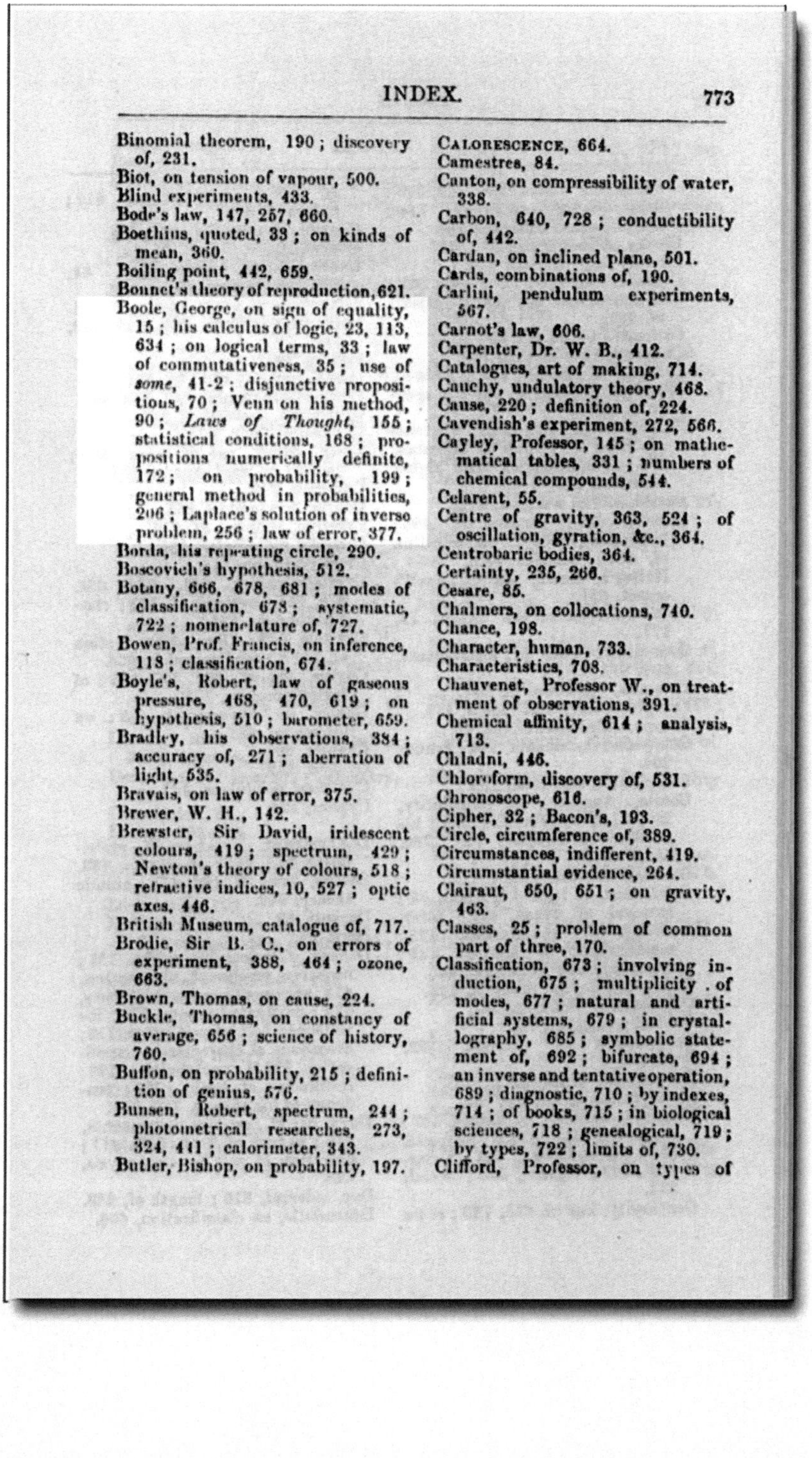

INDEX. 773

Binomial theorem, 190; discovery of, 231.
Biot, on tension of vapour, 500.
Blind experiments, 433.
Bode's law, 147, 257, 660.
Boethius, quoted, 33; on kinds of mean, 360.
Boiling point, 442, 659.
Bonnet's theory of reproduction, 621.
Boole, George, on sign of equality, 15; his calculus of logic, 23, 113, 634; on logical terms, 33; law of commutativeness, 35; use of *some*, 41-2; disjunctive propositions, 70; Venn on his method, 90; *Laws of Thought*, 155; statistical conditions, 168; propositions numerically definite, 172; on probability, 199; general method in probabilities, 206; Laplace's solution of inverse problem, 256; law of error, 377.
Borda, his repeating circle, 290.
Boscovich's hypothesis, 512.
Botany, 666, 678, 681; modes of classification, 678; systematic, 722; nomenclature of, 727.
Bowen, Prof. Francis, on inference, 118; classification, 674.
Boyle's, Robert, law of gaseous pressure, 468, 470, 619; on hypothesis, 510; barometer, 659.
Bradley, his observations, 384; accuracy of, 271; aberration of light, 535.
Bravais, on law of error, 375.
Brewer, W. H., 142.
Brewster, Sir David, iridescent colours, 419; spectrum, 429; Newton's theory of colours, 518; refractive indices, 10, 527; optic axes, 446.
British Museum, catalogue of, 717.
Brodie, Sir B. C., on errors of experiment, 388, 464; ozone, 663.
Brown, Thomas, on cause, 224.
Buckle, Thomas, on constancy of average, 656; science of history, 760.
Buffon, on probability, 215; definition of genius, 576.
Bunsen, Robert, spectrum, 244; photometrical researches, 273, 324, 411; calorimeter, 343.
Butler, Bishop, on probability, 197.
Calorescence, 664.
Camestres, 84.
Canton, on compressibility of water, 338.
Carbon, 640, 728; conductibility of, 442.
Cardan, on inclined plane, 501.
Cards, combinations of, 190.
Carlini, pendulum experiments, 567.
Carnot's law, 606.
Carpenter, Dr. W. B., 412.
Catalogues, art of making, 714.
Cauchy, undulatory theory, 468.
Cause, 220; definition of, 224.
Cavendish's experiment, 272, 566.
Cayley, Professor, 145; on mathematical tables, 331; numbers of chemical compounds, 544.
Celarent, 55.
Centre of gravity, 363, 524; of oscillation, gyration, &c., 364.
Centrobaric bodies, 364.
Certainty, 235, 266.
Cesare, 85.
Chalmers, on collocations, 740.
Chance, 198.
Character, human, 733.
Characteristics, 708.
Chauvenet, Professor W., on treatment of observations, 391.
Chemical affinity, 614; analysis, 713.
Chladni, 446.
Chloroform, discovery of, 531.
Chronoscope, 616.
Cipher, 32; Bacon's, 193.
Circle, circumference of, 389.
Circumstances, indifferent, 419.
Circumstantial evidence, 264.
Clairaut, 650, 651; on gravity, 463.
Classes, 25; problem of common part of three, 170.
Classification, 673; involving induction, 675; multiplicity of modes, 677; natural and artificial systems, 679; in crystallography, 685; symbolic statement of, 692; bifurcate, 694; an inverse and tentative operation, 689; diagnostic, 710; by indexes, 714; of books, 715; in biological sciences, 718; genealogical, 719; by types, 722; limits of, 730.
Clifford, Professor, on types of

Reference by Jevons to De Morgan.

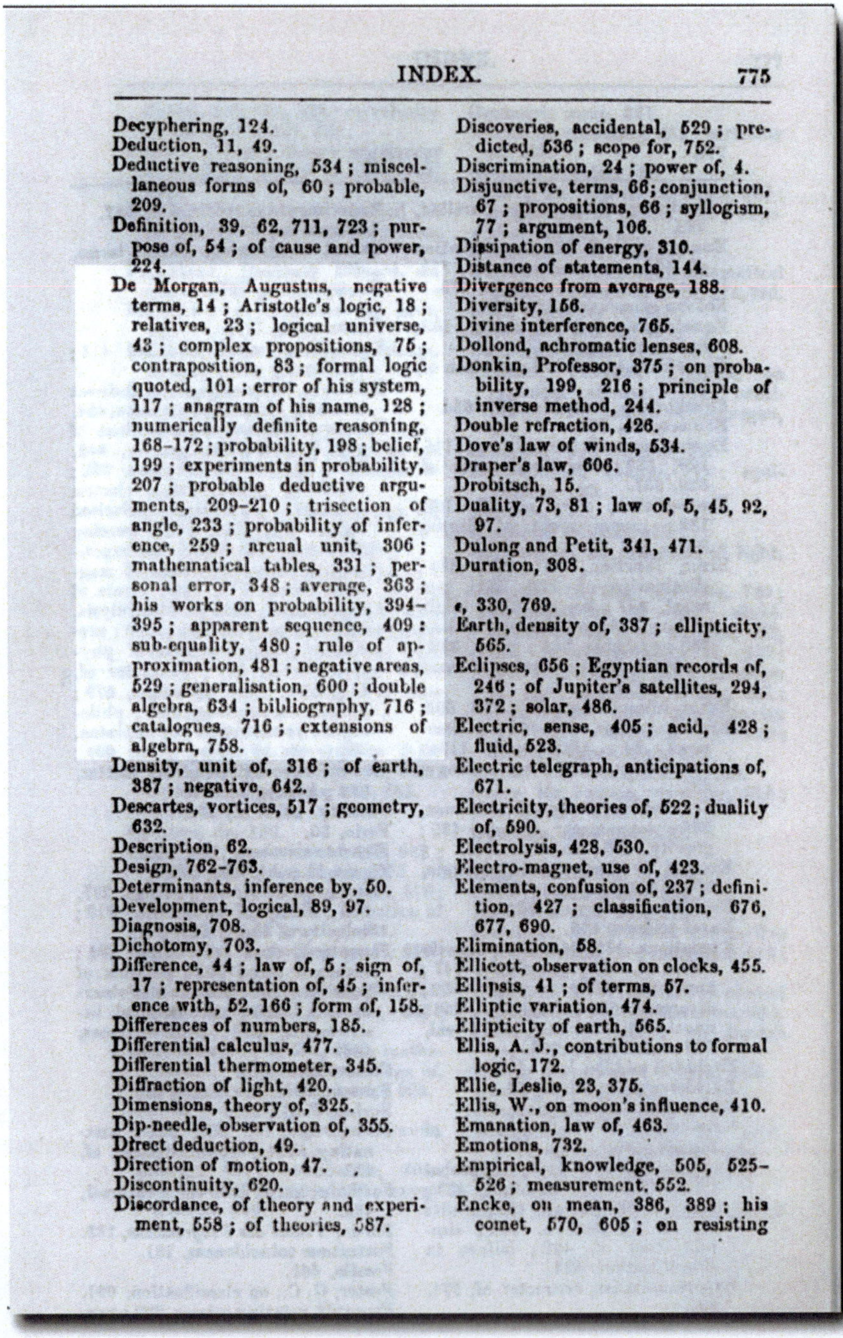

Decyphering, 124.
Deduction, 11, 49.
Deductive reasoning, 534; miscellaneous forms of, 60; probable, 209.
Definition, 39, 62, 711, 723; purpose of, 54; of cause and power, 224.
De Morgan, Augustus, negative terms, 14; Aristotle's logic, 18; relatives, 23; logical universe, 43; complex propositions, 75; contraposition, 83; formal logic quoted, 101; error of his system, 117; anagram of his name, 128; numerically definite reasoning, 168–172; probability, 198; belief, 199; experiments in probability, 207; probable deductive arguments, 209–210; trisection of angle, 233; probability of inference, 259; arcual unit, 306; mathematical tables, 331; personal error, 348; average, 363; his works on probability, 394–395; apparent sequence, 409; sub-equality, 480; rule of approximation, 481; negative areas, 529; generalisation, 600; double algebra, 634; bibliography, 716; catalogues, 716; extensions of algebra, 758.
Density, unit of, 316; of earth, 387; negative, 642.
Descartes, vortices, 517; geometry, 632.
Description, 62.
Design, 762–763.
Determinants, inference by, 50.
Development, logical, 89, 97.
Diagnosis, 708.
Dichotomy, 703.
Difference, 44; law of, 5; sign of, 17; representation of, 45; inference with, 52, 166; form of, 158.
Differences of numbers, 185.
Differential calculus, 477.
Differential thermometer, 345.
Diffraction of light, 420.
Dimensions, theory of, 325.
Dip-needle, observation of, 355.
Direct deduction, 49.
Direction of motion, 47.
Discontinuity, 620.
Discordance, of theory and experiment, 558; of theories, 587.
Discoveries, accidental, 529; predicted, 536; scope for, 752.
Discrimination, 24; power of, 4.
Disjunctive, terms, 66; conjunction, 67; propositions, 66; syllogism, 77; argument, 106.
Dissipation of energy, 310.
Distance of statements, 144.
Divergence from average, 188.
Diversity, 156.
Divine interference, 765.
Dollond, achromatic lenses, 608.
Donkin, Professor, 375; on probability, 199, 216; principle of inverse method, 244.
Double refraction, 426.
Dove's law of winds, 534.
Draper's law, 606.
Drobitsch, 15.
Duality, 73, 81; law of, 5, 45, 92, 97.
Dulong and Petit, 341, 471.
Duration, 308.

e, 330, 769.
Earth, density of, 387; ellipticity, 565.
Eclipses, 656; Egyptian records of, 246; of Jupiter's satellites, 294, 372; solar, 486.
Electric, sense, 405; acid, 428; fluid, 523.
Electric telegraph, anticipations of, 671.
Electricity, theories of, 522; duality of, 590.
Electrolysis, 428, 530.
Electro-magnet, use of, 423.
Elements, confusion of, 237; definition, 427; classification, 676, 677, 690.
Elimination, 58.
Ellicott, observation on clocks, 455.
Ellipsis, 41; of terms, 57.
Elliptic variation, 474.
Ellipticity of earth, 565.
Ellis, A. J., contributions to formal logic, 172.
Ellis, Leslie, 23, 375.
Ellis, W., on moon's influence, 410.
Emanation, law of, 463.
Emotions, 732.
Empirical, knowledge, 505, 525–526; measurement, 552.
Encke, on mean, 386, 389; his comet, 570, 605; on resisting

Platon Sergeevič Poreckij

2.10 Contributions of P.S. Poreckij

Poreckij

1881 Poreckij, P.S., "Referat on basic principles of mathematical logic" in *Protokol tretiago zasedania sekcii fiziko-matematiceskih nauk Obšcestva Estestvoispytatelej pri Imperatorskom Kazanskom Universitete, 17. maja 1880 goda, 230, issued as a part of Protokoly zasedanij Obšcestva Esytestvoispytelej pri Imperatorskom Kazanskom Universitete*, Vol. 12, 1880-1881, Kazan, Russia, 1881.

[Poreckij, Ref. 1]

1884 Poreckij, P.S., "On methods for solving logic equations and inverse by methods of mathematical logic", Kazan

[Poreckij, Ref. 2]

1887 Poreckij, P.S., "Solution of general tasks in probability theory through the mathematical logic", Izd-vo Kazanj Univ, Kazanj, Russia, 1887, printed by the permission of the Society for Natural Sciences at the Imperial University of Kazan, the President of the Society A. Štukenberg.

[Poreckij, Ref. 3]

Work by Poreckij has been referred also in the book *The Algebra of Logic* by L. Couturat.

A strong comment to discuss the subject after Boole.

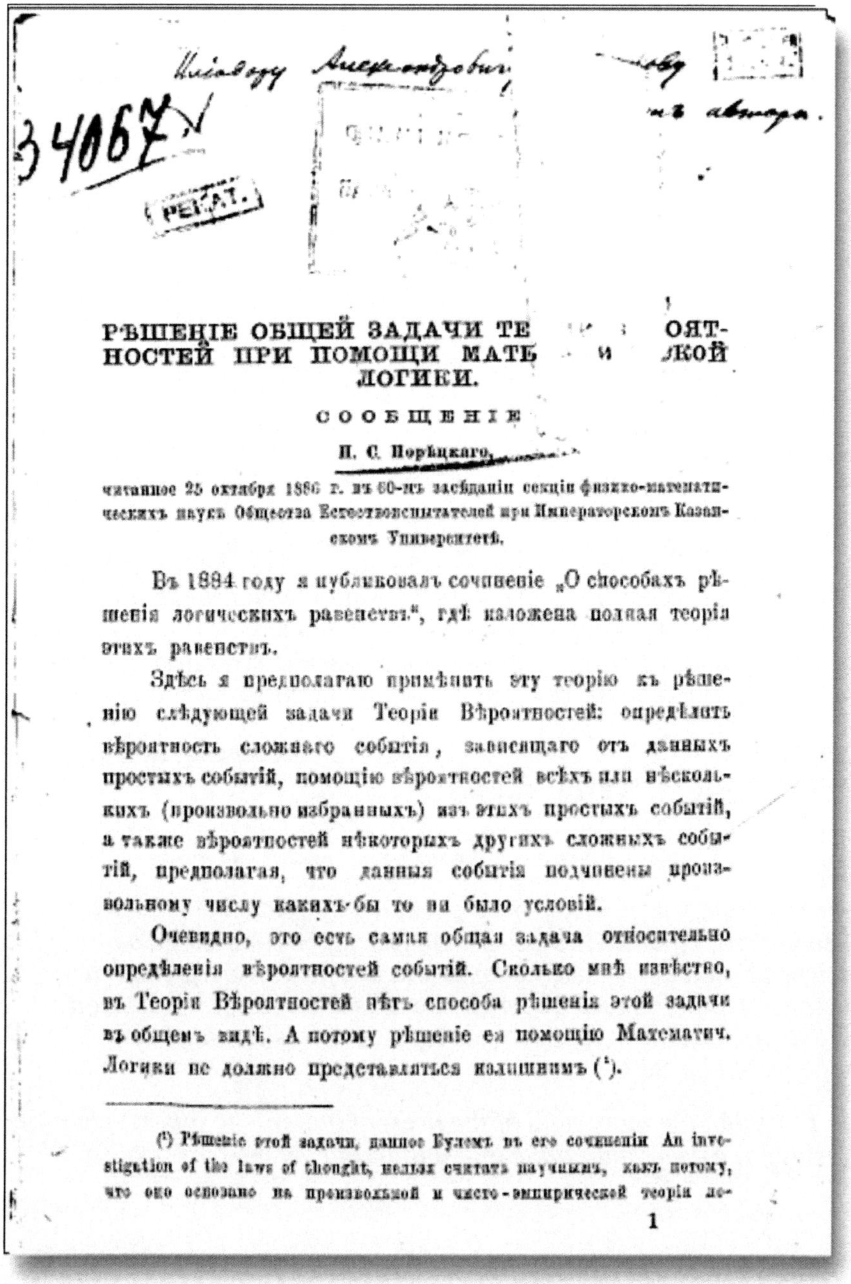

РѢШЕНІЕ ОБЩЕЙ ЗАДАЧИ ТЕ … ОЯТ-НОСТЕЙ ПРИ ПОМОЩИ МАТЕ … КОЙ ЛОГИКИ.

СООБЩЕНІЕ

П. С. Порѣцкаго,

читанное 25 октября 1886 г. въ 60-мъ засѣданіи секціи физико-математическихъ наукъ Общества Естествоиспытателей при Императорскомъ Казанскомъ Университетѣ.

Въ 1884 году я публиковалъ сочиненіе „О способахъ рѣшенія логическихъ равенствъ", гдѣ изложена полная теорія этихъ равенствъ.

Здѣсь я предполагаю примѣнить эту теорію къ рѣшенію слѣдующей задачи Теоріи Вѣроятностей: опредѣлить вѣроятность сложнаго событія, зависящаго отъ данныхъ простыхъ событій, помощію вѣроятностей всѣхъ или нѣкоторыхъ (произвольно избранныхъ) изъ этихъ простыхъ событій, а также вѣроятностей нѣкоторыхъ другихъ сложныхъ событій, предполагая, что данныя событія подчинены произвольному числу какихъ-бы то ни было условій.

Очевидно, это есть самая общая задача относительно опредѣленія вѣроятностей событій. Сколько мнѣ извѣстно, въ Теоріи Вѣроятностей нѣтъ способа рѣшенія этой задачи въ общемъ видѣ. А потому рѣшеніе ея помощію Математич. Логики не должно представляться излишнимъ (¹).

(¹) Рѣшеніе этой задачи, данное Булемъ въ его сочиненіи An investigation of the laws of thought, нельзя считать научнымъ, какъ потому, что оно основано на произвольной и чисто-эмпирической теоріи ло-

1

Motivation for Study of Logic

Poreckij was influenced by the mathematician Alexander V. Vasiliev, father of N.A. Vasiliev, founder of *imaginary (nonaristotelian) logic*

In [Poreckij, Ref.1], there is the first attempt at a complete theory of qualitative inference, where under the term *quality* Poreckij meant *one-place predicate* in modern terminology [Anovskaja], see also [Kline].

Kline, G.L., "Review of *Foundations of Mathematics and Mathematical Logic* by S.A. Anovskaa", *The Journal of Symbolic Logic*, Vol. 16, No. 1, 1951, 46-48.

Anovskaja, S.A., "Osnovania matematiki i matematiceskaa logika (Foundations of mathematics and mathematical logic)", *Mathematika v SSSR za tridcat let 1917 -1947* (Mathematics in the USSR for the thirty years 1917 -1947), OGIZ, Moscow and Leningrad, 1948, 9-50.

In 1884, I published an article "On methods for solving logic equations", where it has been presented a complete theory of these equations. In this article, I suggest to exploit this theory in solving the following task in the Probability Theory.

Determine probability of a complex event, depending on given simple events, by using probabilities of these simple events as well as probabilities of some other complex events, assuming that given events satisfy an arbitrary number of arbitrary conditions.

[Poreckij, Ref. 3]

Poreckij and Boole

In [Poreckij, Ref. 3], Poreckij wrote

A solution of this task has been provided by Boole in his article [Boole, Ref. 1],*which, however, can hardly be considered as scientific, since it is based upon arbitrary and entirely empirical theory of logic, as well as for the idea itself about the transition from logic equation to algebraic equations has weakly been elaborated by Boole.*

In this way, the main goal of the present paper is to give a scientific form to the deep, but vague and without proof, idea of Boole about applicability of Mathematical Logic in the Probability Theory.

Chapter 3
Algebra of Logic

R.S. Stanković & J. Astola: From Boolean Log. to Switch. Circuits & Automata, SCI 335, pp. 87–110.
springerlink.com

Ernst Schröder

A systematic presentation of the Boolean algebra and distributive lattices is given in 1890 by Ernst Schröder.

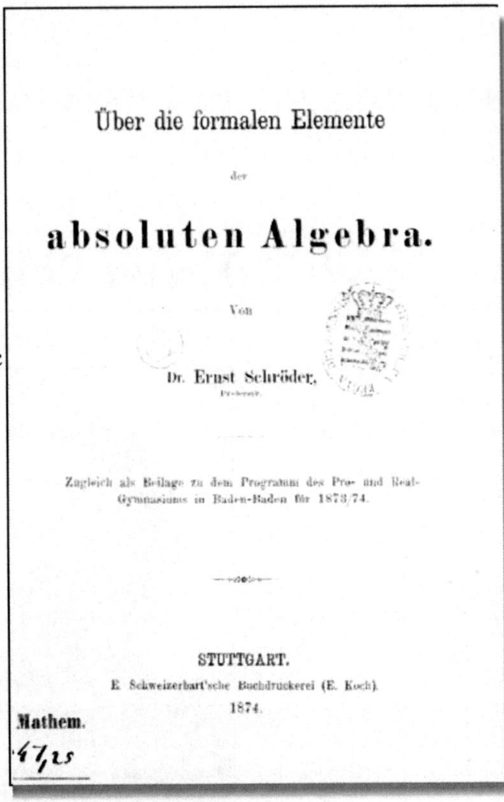

Über die formalen Elemente

der

absoluten Algebra.

Von

Dr. Ernst Schröder,

Professor.

Zugleich als Beilage zu dem Programm des Pro- und Real-Gymnasiums in Baden-Baden für 1873/74.

STUTTGART.

E. Schweizerbart'sche Buchdruckerei (E. Koch).

1874.

Mathem.

47,25

In [Schröder, Ref. 1], Schröder presented a theory of Algebraic logic and this greatly contributed to spreading these ideas in the continental Europe.

Schröder, E., *Der Operationskreis des Logik kalkuls*, B.G. Teubner, Leipzig, Germany, 1877.

It should be noticed that Ernst Schröder developed his algebraic logic, which is called nowadays *symbolic logic*, independently on the work by G.J. Boole and A. De Morgan, about which he learned in 1873. Schröder published his work in [Schröder, Ref. 2], the third volume of which appeared posthumously was edited by Eugen Muller.

[Schröder, Ref. 1]

Schroder, E., *Vorlesungen uber die Algebre der Logik (exacte Logik)*, 1890, 1891, 1895, and 1905, B.G. Teubner, Leipzig, Germany, reprinted in three volumes by Chelsea Publishing Company, Bronx, USA, 1966, and Thoemmes Press, 2000.

[Schröder, Ref. 2]

3.1 Algebra of Logic by Schröder

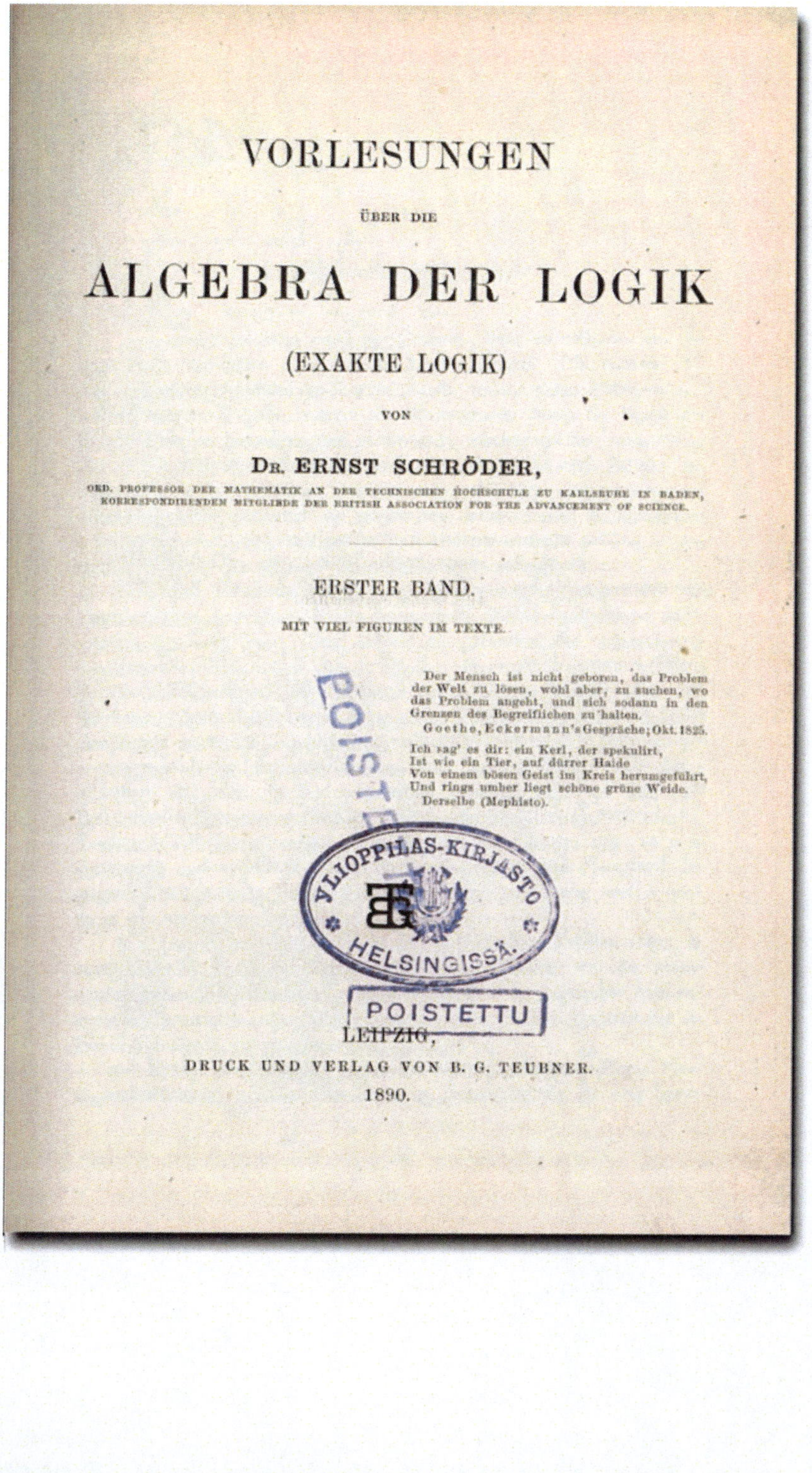
VORLESUNGEN

ÜBER DIE

ALGEBRA DER LOGIK

(EXAKTE LOGIK)

VON

DR. ERNST SCHRÖDER,

ORD. PROFESSOR DER MATHEMATIK AN DER TECHNISCHEN HOCHSCHULE ZU KARLSRUHE IN BADEN,
KORRESPONDIRENDEM MITGLIEDE DER BRITISH ASSOCIATION FOR THE ADVANCEMENT OF SCIENCE.

ERSTER BAND.

MIT VIEL FIGUREN IM TEXTE.

Der Mensch ist nicht geboren, das Problem der Welt zu lösen, wohl aber, zu suchen, wo das Problem angeht, und sich sodann in den Grenzen des Begreiflichen zu halten.
Goethe, Eckermann's Gespräche; Okt. 1825.

Ich sag' es dir: ein Kerl, der spekulirt,
Ist wie ein Tier, auf dürrer Haide
Von einem bösen Geist im Kreis herumgeführt,
Und rings umher liegt schöne grüne Weide.
Derselbe (Mephisto).

LEIPZIG,

DRUCK UND VERLAG VON B. G. TEUBNER.

1890.

Important terms. Exact Logic and Logic Calculus, and a reference to Charles Peirce.

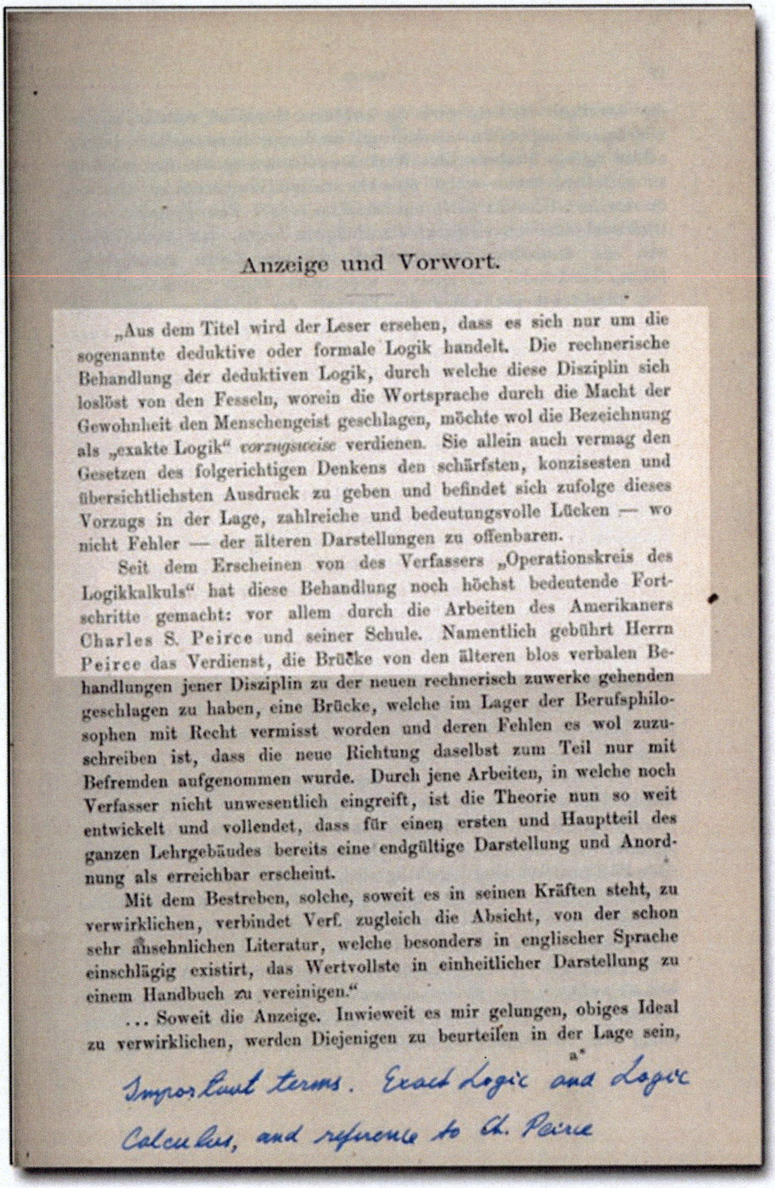

Anzeige und Vorwort.

„Aus dem Titel wird der Leser ersehen, dass es sich nur um die sogenannte deduktive oder formale Logik handelt. Die rechnerische Behandlung der deduktiven Logik, durch welche diese Disziplin sich loslöst von den Fesseln, worein die Wortsprache durch die Macht der Gewohnheit den Menschengeist geschlagen, möchte wol die Bezeichnung als „exakte Logik“ *vorzugsweise* verdienen. Sie allein auch vermag den Gesetzen des folgerichtigen Denkens den schärfsten, konzisesten und übersichtlichsten Ausdruck zu geben und befindet sich zufolge dieses Vorzugs in der Lage, zahlreiche und bedeutungsvolle Lücken — wo nicht Fehler — der älteren Darstellungen zu offenbaren.

Seit dem Erscheinen von des Verfassers „Operationskreis des Logikkalkuls“ hat diese Behandlung noch höchst bedeutende Fortschritte gemacht: vor allem durch die Arbeiten des Amerikaners Charles S. Peirce und seiner Schule. Namentlich gebührt Herrn Peirce das Verdienst, die Brücke von den älteren blos verbalen Behandlungen jener Disziplin zu der neuen rechnerisch zuwerke gehenden geschlagen zu haben, eine Brücke, welche im Lager der Berufsphilosophen mit Recht vermisst worden und deren Fehlen es wol zuzuschreiben ist, dass die neue Richtung daselbst zum Teil nur mit Befremden aufgenommen wurde. Durch jene Arbeiten, in welche noch Verfasser nicht unwesentlich eingreift, ist die Theorie nun so weit entwickelt und vollendet, dass für einen ersten und Hauptteil des ganzen Lehrgebäudes bereits eine endgültige Darstellung und Anordnung als erreichbar erscheint.

Mit dem Bestreben, solche, soweit es in seinen Kräften steht, zu verwirklichen, verbindet Verf. zugleich die Absicht, von der schon sehr ansehnlichen Literatur, welche besonders in englischer Sprache einschlägig existirt, das Wertvollste in einheitlicher Darstellung zu einem Handbuch zu vereinigen.“

... Soweit die Anzeige. Inwieweit es mir gelungen, obiges Ideal zu verwirklichen, werden Diejenigen zu beurteilen in der Lage sein,

a*

Important terms. Exact Logic and Logic Calculus, and reference to Ch. Peirce

Reference to Mill, Stuart, and Sigwart.

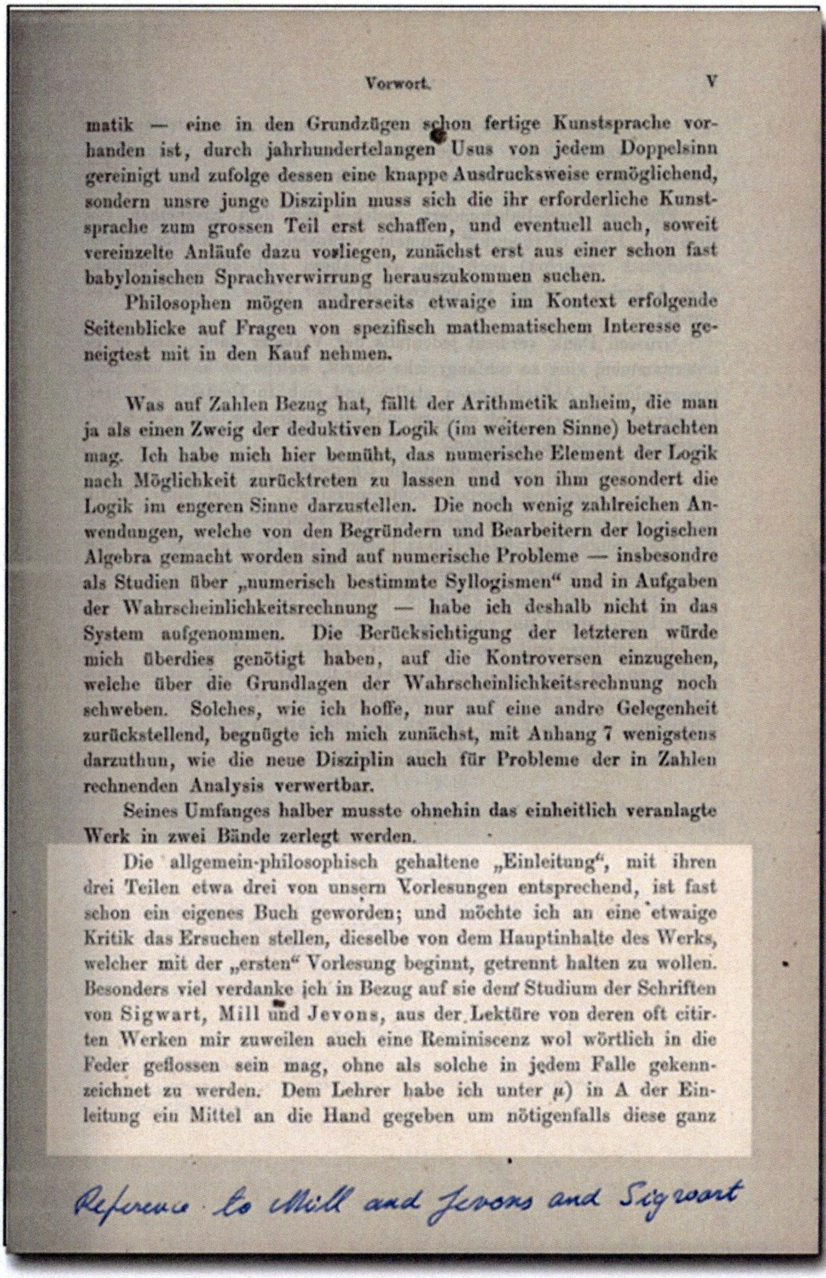

Vorwort. V

matik — eine in den Grundzügen schon fertige Kunstsprache vorhanden ist, durch jahrhundertelangen Usus von jedem Doppelsinn gereinigt und zufolge dessen eine knappe Ausdrucksweise ermöglichend, sondern unsre junge Disziplin muss sich die ihr erforderliche Kunstsprache zum grossen Teil erst schaffen, und eventuell auch, soweit vereinzelte Anläufe dazu vorliegen, zunächst erst aus einer schon fast babylonischen Sprachverwirrung herauszukommen suchen.

Philosophen mögen andrerseits etwaige im Kontext erfolgende Seitenblicke auf Fragen von spezifisch mathematischem Interesse geneigtest mit in den Kauf nehmen.

Was auf Zahlen Bezug hat, fällt der Arithmetik anheim, die man ja als einen Zweig der deduktiven Logik (im weiteren Sinne) betrachten mag. Ich habe mich hier bemüht, das numerische Element der Logik nach Möglichkeit zurücktreten zu lassen und von ihm gesondert die Logik im engeren Sinne darzustellen. Die noch wenig zahlreichen Anwendungen, welche von den Begründern und Bearbeitern der logischen Algebra gemacht worden sind auf numerische Probleme — insbesondre als Studien über „numerisch bestimmte Syllogismen“ und in Aufgaben der Wahrscheinlichkeitsrechnung — habe ich deshalb nicht in das System aufgenommen. Die Berücksichtigung der letzteren würde mich überdies genötigt haben, auf die Kontroversen einzugehen, welche über die Grundlagen der Wahrscheinlichkeitsrechnung noch schweben. Solches, wie ich hoffe, nur auf eine andre Gelegenheit zurückstellend, begnügte ich mich zunächst, mit Anhang 7 wenigstens darzuthun, wie die neue Disziplin auch für Probleme der in Zahlen rechnenden Analysis verwertbar.

Seines Umfanges halber musste ohnehin das einheitlich veranlagte Werk in zwei Bände zerlegt werden.

Die allgemein-philosophisch gehaltene „Einleitung“, mit ihren drei Teilen etwa drei von unsern Vorlesungen entsprechend, ist fast schon ein eigenes Buch geworden; und möchte ich an eine etwaige Kritik das Ersuchen stellen, dieselbe von dem Hauptinhalte des Werks, welcher mit der „ersten“ Vorlesung beginnt, getrennt halten zu wollen. Besonders viel verdanke ich in Bezug auf sie dem Studium der Schriften von Sigwart, Mill und Jevons, aus der Lektüre von deren oft citirten Werken mir zuweilen auch eine Reminiscenz wol wörtlich in die Feder geflossen sein mag, ohne als solche in jedem Falle gekennzeichnet zu werden. Dem Lehrer habe ich unter μ) in A der Einleitung ein Mittel an die Hand gegeben um nötigenfalls diese ganz

Reference to Mill and Jevons and Sigwart

Reference to the *Laws of Thought* by Boole.

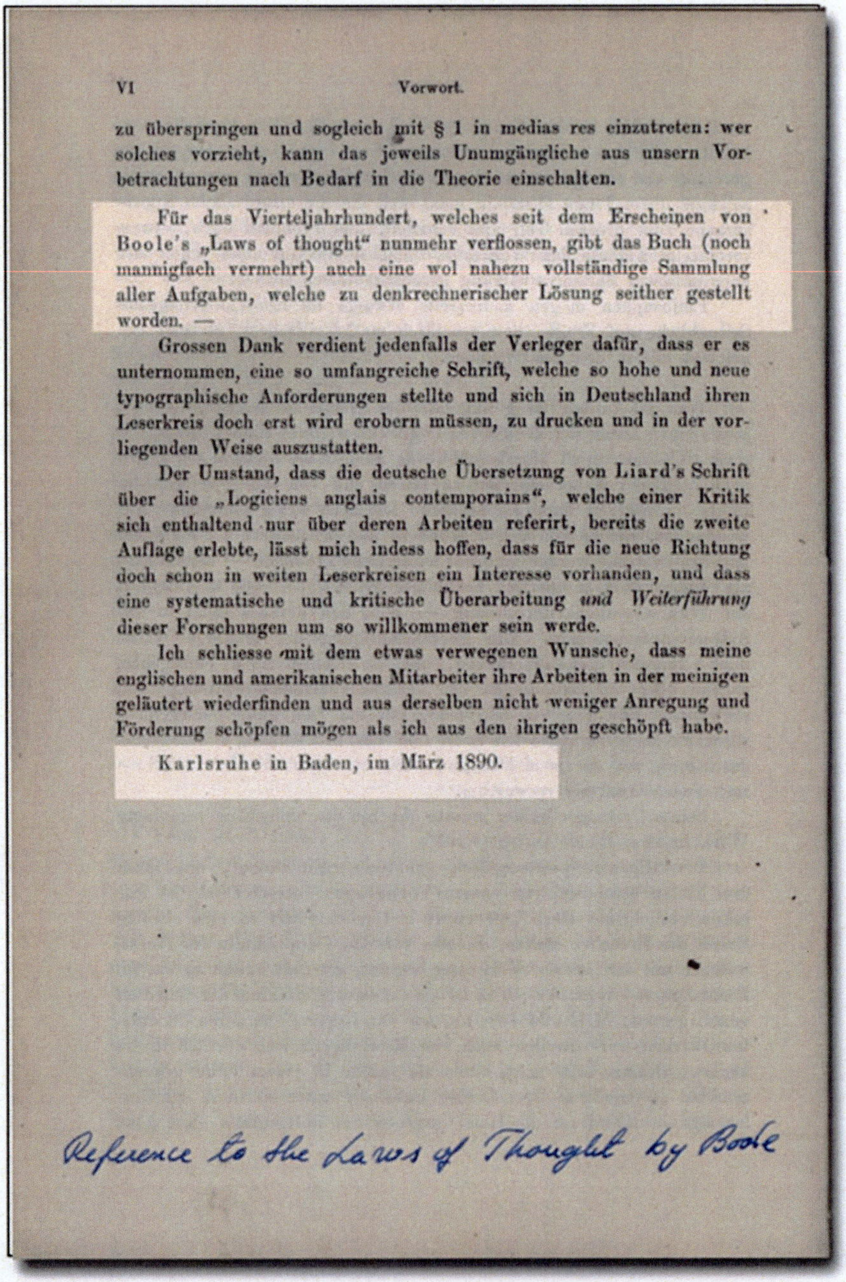

VI Vorwort.

zu überspringen und sogleich mit § 1 in medias res einzutreten: wer solches vorzieht, kann das jeweils Unumgängliche aus unsern Vorbetrachtungen nach Bedarf in die Theorie einschalten.

Für das Vierteljahrhundert, welches seit dem Erscheinen von Boole's „Laws of thought" nunmehr verflossen, gibt das Buch (noch mannigfach vermehrt) auch eine wol nahezu vollständige Sammlung aller Aufgaben, welche zu denkrechnerischer Lösung seither gestellt worden. —

Grossen Dank verdient jedenfalls der Verleger dafür, dass er es unternommen, eine so umfangreiche Schrift, welche so hohe und neue typographische Anforderungen stellte und sich in Deutschland ihren Leserkreis doch erst wird erobern müssen, zu drucken und in der vorliegenden Weise auszustatten.

Der Umstand, dass die deutsche Übersetzung von Liard's Schrift über die „Logiciens anglais contemporains", welche einer Kritik sich enthaltend nur über deren Arbeiten referirt, bereits die zweite Auflage erlebte, lässt mich indess hoffen, dass für die neue Richtung doch schon in weiten Leserkreisen ein Interesse vorhanden, und dass eine systematische und kritische Überarbeitung *und Weiterführung* dieser Forschungen um so willkommener sein werde.

Ich schliesse mit dem etwas verwegenen Wunsche, dass meine englischen und amerikanischen Mitarbeiter ihre Arbeiten in der meinigen geläutert wiederfinden und aus derselben nicht weniger Anregung und Förderung schöpfen mögen als ich aus den ihrigen geschöpft habe.

Karlsruhe in Baden, im März 1890.

Reference to the Laws of Thought by Boole

Reference to Boole and Carol Lewis (Charles Lutwidge Dodgson).

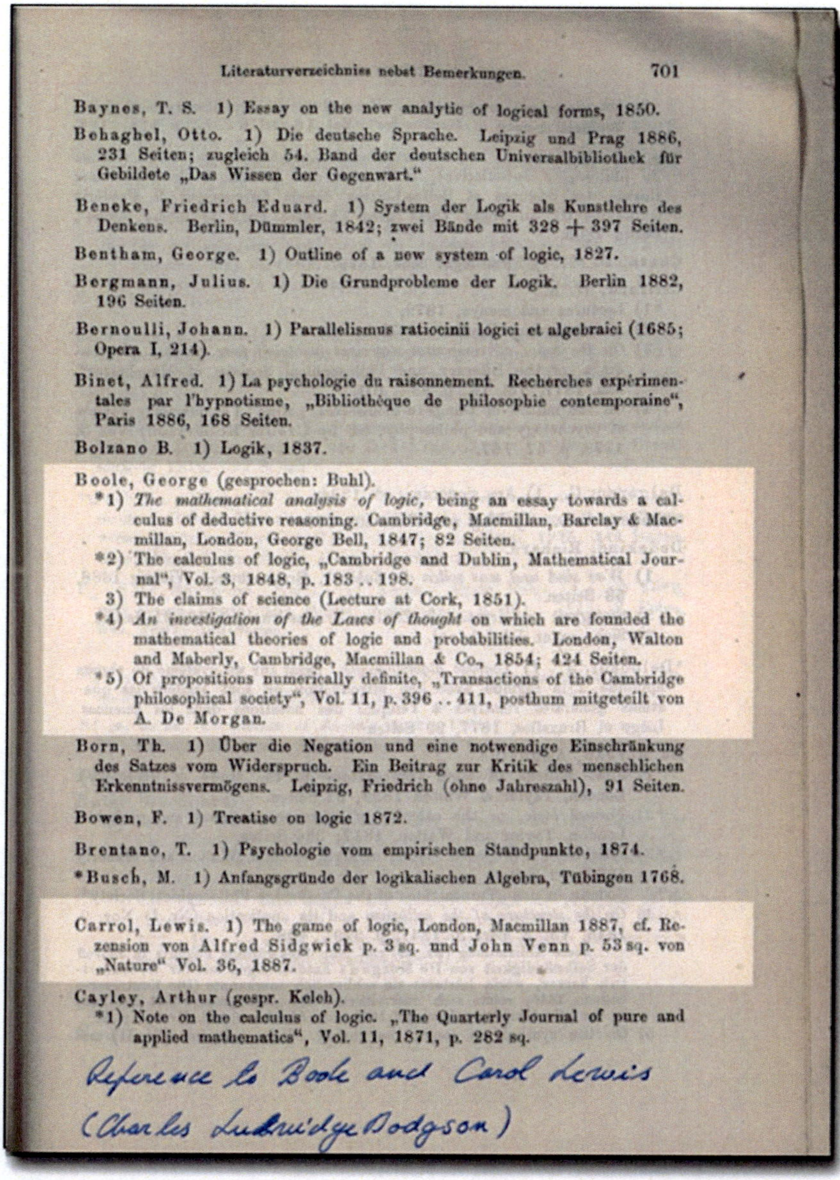

Literaturverzeichniss nebst Bemerkungen. 701

Baynes, T. S. 1) Essay on the new analytic of logical forms, 1850.

Behaghel, Otto. 1) Die deutsche Sprache. Leipzig und Prag 1886, 231 Seiten; zugleich 54. Band der deutschen Universalbibliothek für Gebildete „Das Wissen der Gegenwart."

Beneke, Friedrich Eduard. 1) System der Logik als Kunstlehre des Denkens. Berlin, Dümmler, 1842; zwei Bände mit 328 + 397 Seiten.

Bentham, George. 1) Outline of a new system of logic, 1827.

Bergmann, Julius. 1) Die Grundprobleme der Logik. Berlin 1882, 196 Seiten.

Bernoulli, Johann. 1) Parallelismus ratiocinii logici et algebraici (1685; Opera I, 214).

Binet, Alfred. 1) La psychologie du raisonnement. Recherches expérimentales par l'hypnotisme, „Bibliothèque de philosophie contemporaine", Paris 1886, 168 Seiten.

Bolzano B. 1) Logik, 1837.

Boole, George (gesprochen: Buhl).
*1) *The mathematical analysis of logic*, being an essay towards a calculus of deductive reasoning. Cambridge, Macmillan, Barclay & Macmillan, London, George Bell, 1847; 82 Seiten.
*2) The calculus of logic, „Cambridge and Dublin, Mathematical Journal", Vol. 3, 1848, p. 183 .. 198.
3) The claims of science (Lecture at Cork, 1851).
*4) *An investigation of the Laws of thought* on which are founded the mathematical theories of logic and probabilities. London, Walton and Maberly, Cambridge, Macmillan & Co., 1854; 424 Seiten.
*5) Of propositions numerically definite, „Transactions of the Cambridge philosophical society", Vol. 11, p. 396 .. 411, posthum mitgeteilt von A. De Morgan.

Born, Th. 1) Über die Negation und eine notwendige Einschränkung des Satzes vom Widerspruch. Ein Beitrag zur Kritik des menschlichen Erkenntnissvermögens. Leipzig, Friedrich (ohne Jahreszahl), 91 Seiten.

Bowen, F. 1) Treatise on logic 1872.

Brentano, T. 1) Psychologie vom empirischen Standpunkte, 1874.

*Busch, M. 1) Anfangsgründe der logikalischen Algebra, Tübingen 1768.

Carrol, Lewis. 1) The game of logic, London, Macmillan 1887, cf. Rezension von Alfred Sidgwick p. 3 sq. und John Venn p. 53 sq. von „Nature" Vol. 36, 1887.

Cayley, Arthur (gespr. Kelch).
*1) Note on the calculus of logic. „The Quarterly Journal of pure and applied mathematics", Vol. 11, 1871, p. 282 sq.

Reference to Boole and Carol Lewis (Charles Ludwidge Dodgson)

Reference to De Morgan.

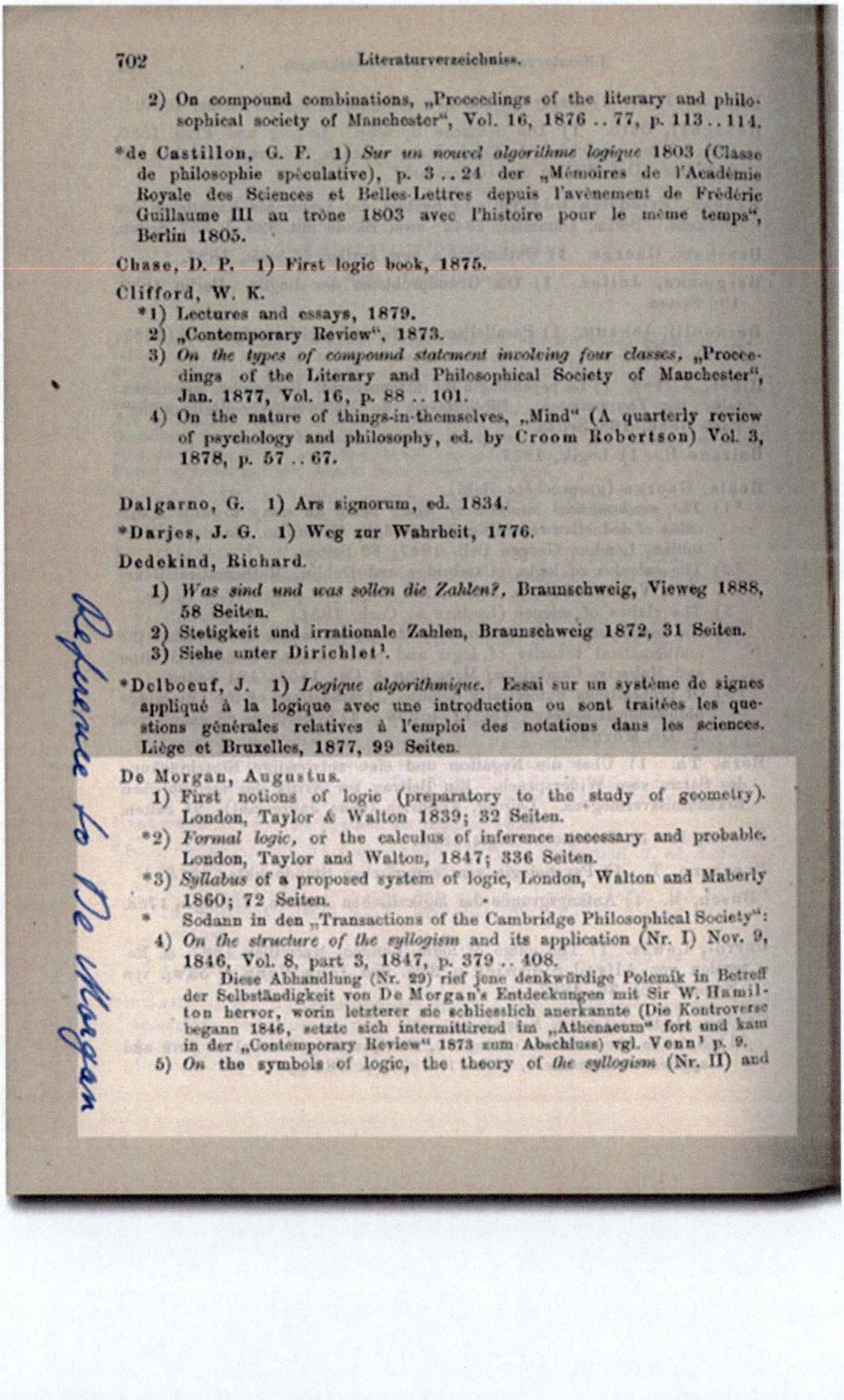

702 Literaturverzeichniss.

2) On compound combinations, „Proceedings of the literary and philosophical society of Manchester", Vol. 16, 1876 .. 77, p. 113 .. 114.

*de Castillon, G. F. 1) *Sur un nouvel algorithme logique* 1803 (Classe de philosophie spéculative), p. 3 .. 24 der „Mémoires de l'Académie Royale des Sciences et Belles-Lettres depuis l'avènement de Frédéric Guillaume III au trône 1803 avec l'histoire pour le même temps", Berlin 1805.

Chase, D. P. 1) First logic book, 1875.

Clifford, W. K.

*1) Lectures and essays, 1879.

2) „Contemporary Review", 1873.

3) *On the types of compound statement involving four classes*, „Proceedings of the Literary and Philosophical Society of Manchester", Jan. 1877, Vol. 16, p. 88 .. 101.

4) On the nature of things-in-themselves, „Mind" (A quarterly review of psychology and philosophy, ed. by Croom Robertson) Vol. 3, 1878, p. 57 .. 67.

Dalgarno, G. 1) Ars signorum, ed. 1834.

*Darjes, J. G. 1) Weg zur Wahrheit, 1776.

Dedekind, Richard.

1) *Was sind und was sollen die Zahlen?*, Braunschweig, Vieweg 1888, 58 Seiten.

2) Stetigkeit und irrationale Zahlen, Braunschweig 1872, 31 Seiten.

3) Siehe unter Dirichlet[1].

*Delboeuf, J. 1) *Logique algorithmique*. Essai sur un système de signes appliqué à la logique avec une introduction ou sont traitées les questions générales relatives à l'emploi des notations dans les sciences. Liège et Bruxelles, 1877, 99 Seiten.

De Morgan, Augustus.

1) First notions of logic (preparatory to the study of geometry). London, Taylor & Walton 1839; 32 Seiten.

*2) *Formal logic*, or the calculus of inference necessary and probable. London, Taylor and Walton, 1847; 336 Seiten.

*3) *Syllabus* of a proposed system of logic, London, Walton and Maberly 1860; 72 Seiten.

* Sodann in den „Transactions of the Cambridge Philosophical Society":

4) *On the structure of the syllogism* and its application (Nr. I) Nov. 9, 1846, Vol. 8, part 3, 1847, p. 379 .. 408.

Diese Abhandlung (Nr. 29) rief jene denkwürdige Polemik in Betreff der Selbständigkeit von De Morgan's Entdeckungen mit Sir W. Hamilton hervor, worin letzterer sie schliesslich anerkannte (Die Kontroverse begann 1846, setzte sich intermittirend im „Athenaeum" fort und kam in der „Contemporary Review" 1873 zum Abschluss) vgl. Venn[1] p. 9.

5) *On* the symbols of logic, the theory of *the syllogism* (Nr. II) and

in particular of the copula, and the application of the theory of probabilities to some questions of evidence. Febr. 25, 1850, Vol. 9, 1851, part. 1, p. 79 .. 127.

6) *On the syllogism* Nr. III and on logic in general. Febr. 8, 1858, Vol. 10, 1864, part. 1, p. 173 .. 230.

7) *On the syllogism* Nr. IV and on the logic of relations. April 23, 1860, Vol. 10, 1864, part. 2, p. 331 .. 358.

8) *On the syllogism* Nr. V and on various points of the onymatic system. May 4, 1863, Vol. 10, 1864, part. 2, p. 428 .. 487.

9) Artikel *Logic* in der „Englisch Cyclopaedia" von 1860.

10) Vergl. Boole[5].

Dieffenbach, Ludwig. 1) Der menschliche Wille und seine Grundlagen. Die Freiheit des Willens und die Zurechnung. Darmstadt 1886, 130 S. (Selbstverlag des nun verstorbenen Verfassers, C. F. Winter'sche Buchdruckerei.)

Dirichlet, Lejeune. 1) Vorlesungen über Zahlentheorie, herausgegeben und mit Zusätzen versehen von **Dedekind**, 3. Aufl., 2 Bände. Braunschweig 1879, 627 Seiten.

*__Drobisch, Moritz Wilhelm.__ 1) Neue Darstellung der Logik nach ihren einfachsten Verhältnissen mit Rücksicht auf Mathematik und Naturwissenschaften. Vierte Auflage. Leipzig, L. Voss, 1875, 244 Seiten. Inzwischen ist eine fünfte Auflage erschienen.

Du Bois Reymond, Emil. 1) Reden. Erste Folge, Leipzig 1886, 550 Seiten (darin: Die sieben Welträthsel, p. 381 .. 411), 2. Folge, ibid. 1887, 589 Seiten.

*__Ellis, A., J.__ 1) On the algebraical analogues of logical relations, „Proceedings of the Royal Society of London", Vol. 21, p. 497 sq.

Ellis, R. L. 1) Edition of Bacon's works, 1858.
*2) Mathematical and other writings 1863.

Erdmann, J. E. 1) Geschichte der neueren Philosophie 1834 .. 53.

Euler, Leonhard.
1) Briefe an eine deutsche Fürstinn über verschiedene Gegenstände aus der Naturlehre. Nach der Ausgabe von **Condorcet** und **de la Croix**, übers. von **F. Kries**. Leipzig 1792 .. 94, 3 Bde. von 547 + 384 + 424 Seiten. Das Original führt den Titel: Lettres à une princesse d'Allemagne sur quelque sujets de physique et de philosophie, 1768 .. 72 — daselbst vergleiche II p. 106, Lettre 102 .. 105 — auch existirt eine englische Ausgabe: Letters to a German Princess, Ed. Brewster, 1823.

Franklin, Frau, s. **Ladd**.

Frege, Gottlob.

Reference to Jevons.

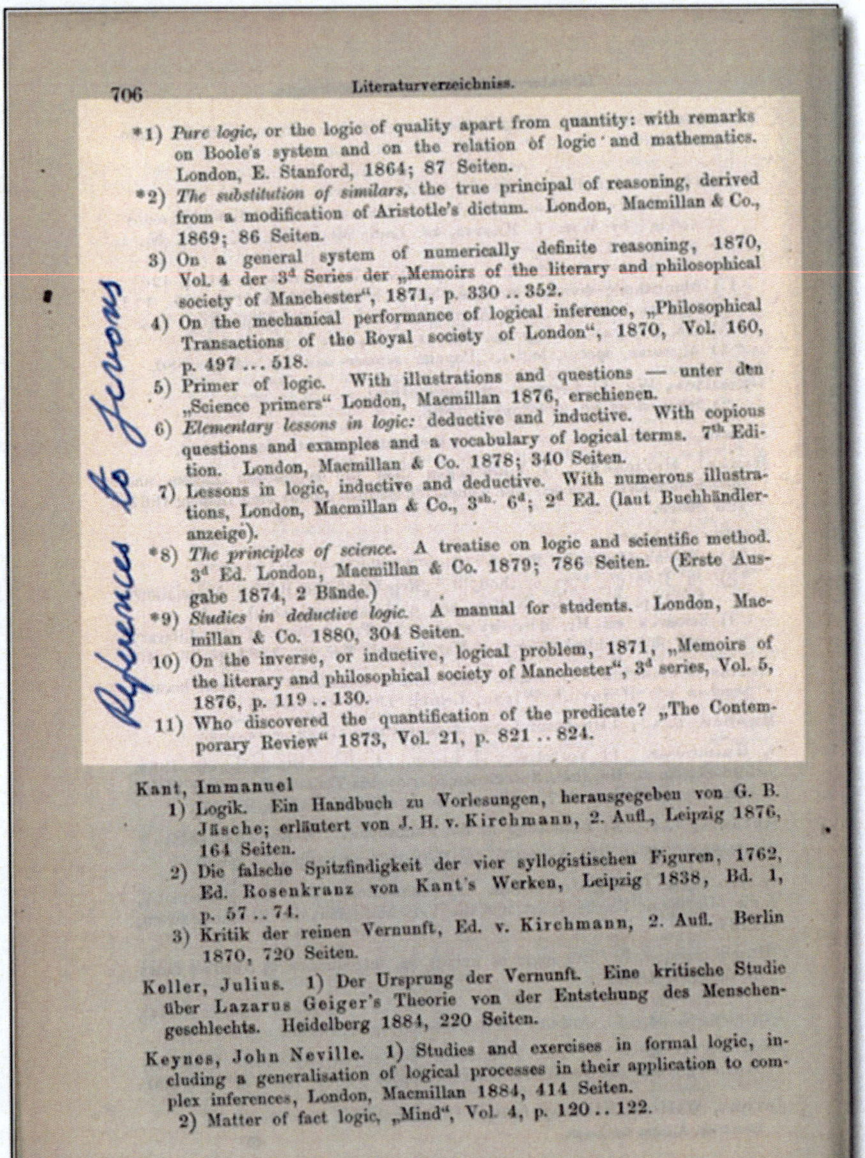

706 Literaturverzeichniss.

*1) *Pure logic,* or the logic of quality apart from quantity: with remarks on Boole's system and on the relation of logic and mathematics. London, E. Stanford, 1864; 87 Seiten.

*2) *The substitution of similars,* the true principal of reasoning, derived from a modification of Aristotle's dictum. London, Macmillan & Co., 1869; 86 Seiten.

3) On a general system of numerically definite reasoning, 1870, Vol. 4 der 3^{d} Series der „Memoirs of the literary and philosophical society of Manchester", 1871, p. 330 .. 352.

4) On the mechanical performance of logical inference, „Philosophical Transactions of the Royal society of London", 1870, Vol. 160, p. 497 ... 518.

5) Primer of logic. With illustrations and questions — unter den „Science primers" London, Macmillan 1876, erschienen.

6) *Elementary lessons in logic:* deductive and inductive. With copious questions and examples and a vocabulary of logical terms. 7th Edition. London, Macmillan & Co. 1878; 340 Seiten.

7) Lessons in logic, inductive and deductive. With numerous illustrations, London, Macmillan & Co., $3^{sh.}$ 6^{d}; 2^{d} Ed. (laut Buchhändleranzeige).

*8) *The principles of science.* A treatise on logic and scientific method. 3^{d} Ed. London, Macmillan & Co. 1879; 786 Seiten. (Erste Ausgabe 1874, 2 Bände.)

*9) *Studies in deductive logic.* A manual for students. London, Macmillan & Co. 1880, 304 Seiten.

10) On the inverse, or inductive, logical problem, 1871, „Memoirs of the literary and philosophical society of Manchester", 3^{d} series, Vol. 5, 1876, p. 119 .. 130.

11) Who discovered the quantification of the predicate? „The Contemporary Review" 1873, Vol. 21, p. 821 .. 824.

Kant, Immanuel

1) Logik. Ein Handbuch zu Vorlesungen, herausgegeben von G. B. Jäsche; erläutert von J. H. v. Kirchmann, 2. Aufl., Leipzig 1876, 164 Seiten.

2) Die falsche Spitzfindigkeit der vier syllogistischen Figuren, 1762, Ed. Rosenkranz von Kant's Werken, Leipzig 1838, Bd. 1, p. 57 .. 74.

3) Kritik der reinen Vernunft, Ed. v. Kirchmann, 2. Aufl. Berlin 1870, 720 Seiten.

Keller, Julius. 1) Der Ursprung der Vernunft. Eine kritische Studie über Lazarus Geiger's Theorie von der Entstehung des Menschengeschlechts. Heidelberg 1884, 220 Seiten.

Keynes, John Neville. 1) Studies and exercises in formal logic, including a generalisation of logical processes in their application to complex inferences, London, Macmillan 1884, 414 Seiten.

2) Matter of fact logic, „Mind", Vol. 4, p. 120 .. 122.

Reference to the John Stuart Mill.

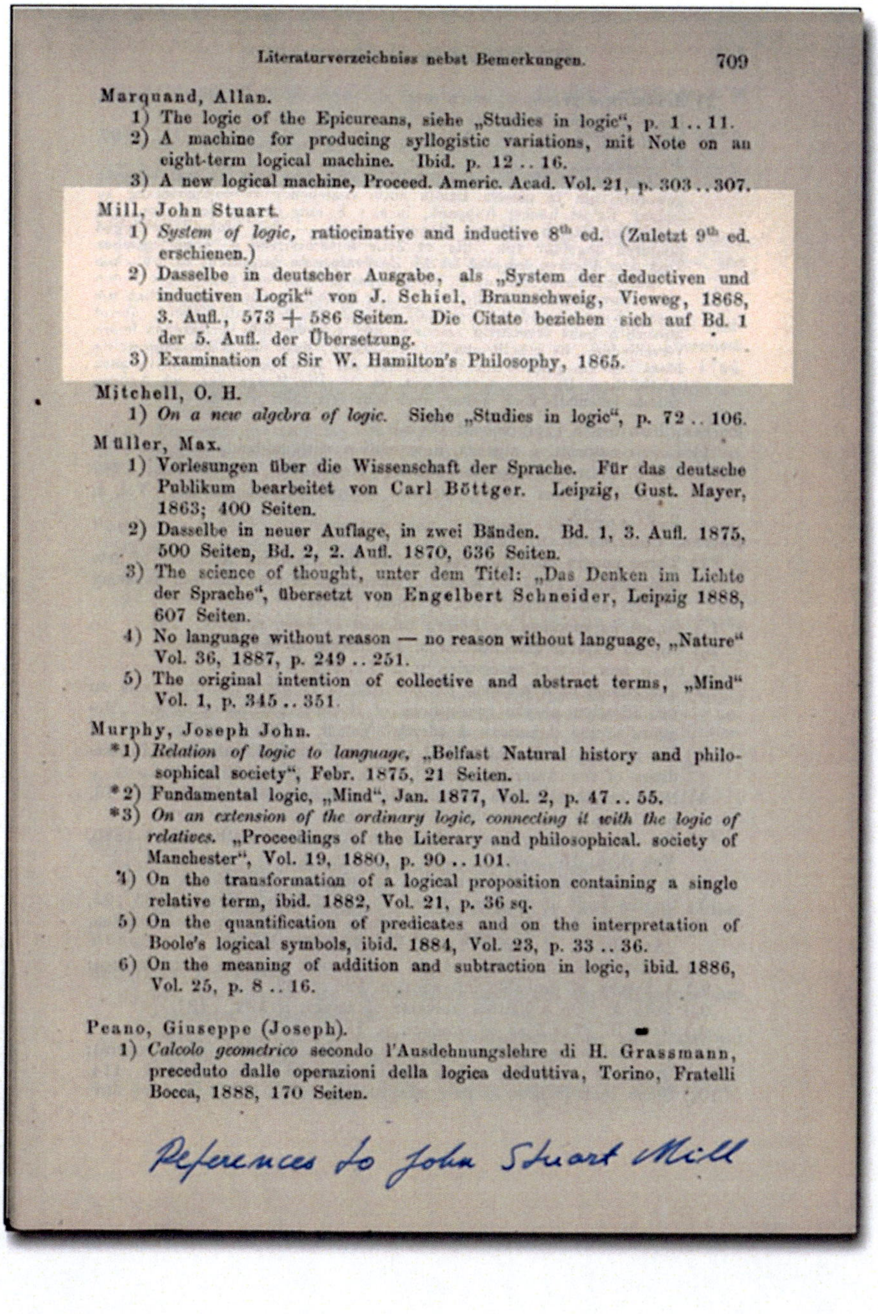

Literaturverzeichniss nebst Bemerkungen. 709

Marquand, Allan.

1) The logic of the Epicureans, siehe „Studies in logic", p. 1 .. 11.
2) A machine for producing syllogistic variations, mit Note on an eight-term logical machine. Ibid. p. 12 .. 16.
3) A new logical machine, Proceed. Americ. Acad. Vol. 21, p. 303 .. 307.

Mill, John Stuart.

1) *System of logic,* ratiocinative and inductive 8th ed. (Zuletzt 9th ed. erschienen.)
2) Dasselbe in deutscher Ausgabe, als „System der deductiven und inductiven Logik" von J. Schiel, Braunschweig, Vieweg, 1868, 3. Aufl., 573 + 586 Seiten. Die Citate beziehen sich auf Bd. 1 der 5. Aufl. der Übersetzung.
3) Examination of Sir W. Hamilton's Philosophy, 1865.

Mitchell, O. H.

1) *On a new algebra of logic.* Siehe „Studies in logic", p. 72 .. 106.

Müller, Max.

1) Vorlesungen über die Wissenschaft der Sprache. Für das deutsche Publikum bearbeitet von Carl Böttger. Leipzig, Gust. Mayer, 1863; 400 Seiten.
2) Dasselbe in neuer Auflage, in zwei Bänden. Bd. 1, 3. Aufl. 1875, 500 Seiten, Bd. 2, 2. Aufl. 1870, 636 Seiten.
3) The science of thought, unter dem Titel: „Das Denken im Lichte der Sprache", übersetzt von Engelbert Schneider, Leipzig 1888, 607 Seiten.
4) No language without reason — no reason without language, „Nature" Vol. 36, 1887, p. 249 .. 251.
5) The original intention of collective and abstract terms, „Mind" Vol. 1, p. 345 .. 351.

Murphy, Joseph John.

*1) *Relation of logic to language,* „Belfast Natural history and philosophical society", Febr. 1875, 21 Seiten.
*2) Fundamental logic, „Mind", Jan. 1877, Vol. 2, p. 47 .. 55.
*3) *On an extension of the ordinary logic, connecting it with the logic of relatives.* „Proceedings of the Literary and philosophical. society of Manchester", Vol. 19, 1880, p. 90 .. 101.
4) On the transformation of a logical proposition containing a single relative term, ibid. 1882, Vol. 21, p. 36 sq.
5) On the quantification of predicates and on the interpretation of Boole's logical symbols, ibid. 1884, Vol. 23, p. 33 .. 36.
6) On the meaning of addition and subtraction in logic, ibid. 1886, Vol. 25, p. 8 .. 16.

Peano, Giuseppe (Joseph).

1) *Calcolo geometrico* secondo l'Ausdehnungslehre di H. Grassmann, preceduto dalle operazioni della logica deduttiva, Torino, Fratelli Bocca, 1888, 170 Seiten.

References to John Stuart Mill

Reference to Charles Peirce.

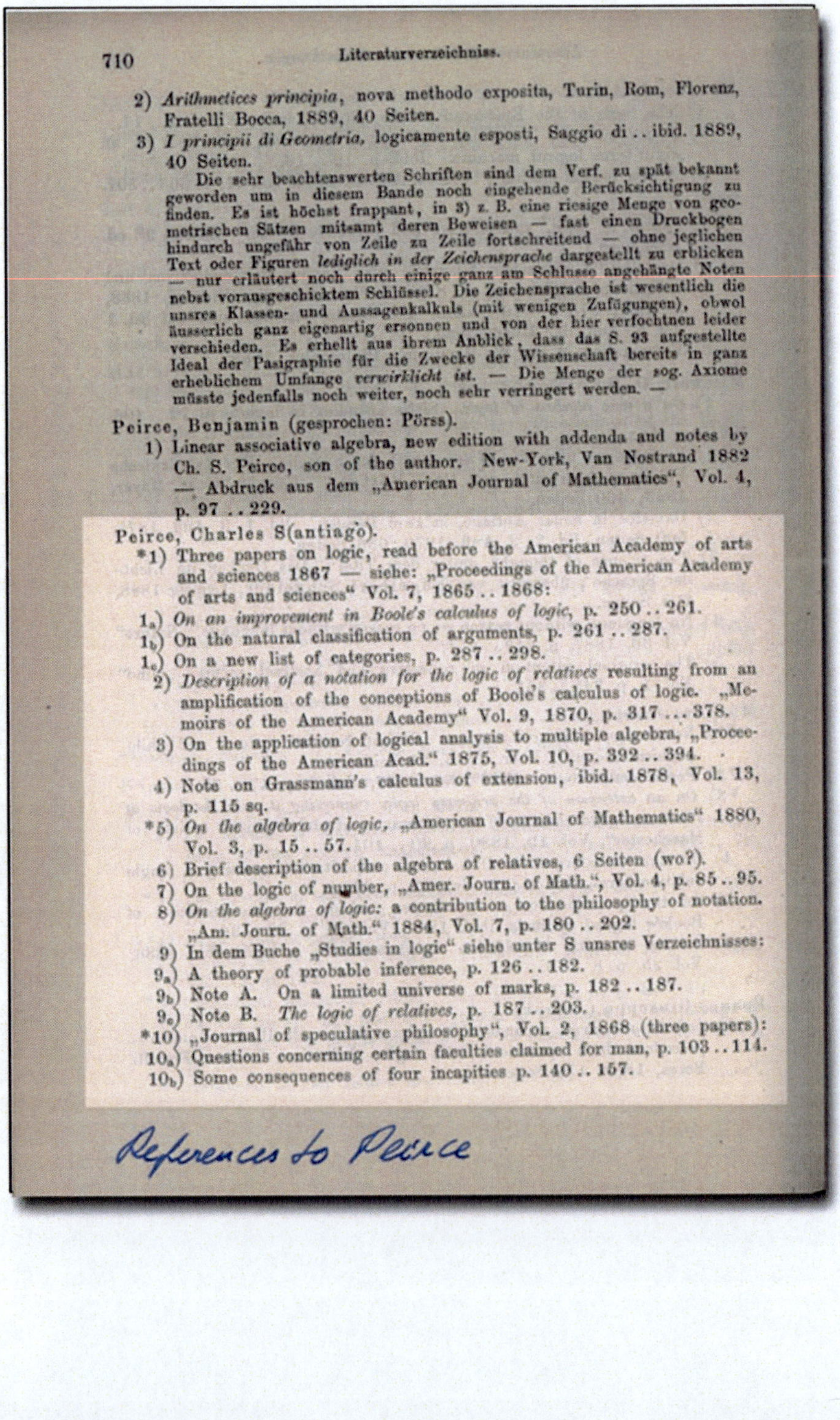

710 Literaturverzeichniss.

2) *Arithmetices principia*, nova methodo exposita, Turin, Rom, Florenz, Fratelli Bocca, 1889, 40 Seiten.

3) *I principii di Geometria,* logicamente esposti, Saggio di .. ibid. 1889, 40 Seiten.

Die sehr beachtenswerten Schriften sind dem Verf. zu spät bekannt geworden um in diesem Bande noch eingehende Berücksichtigung zu finden. Es ist höchst frappant, in 3) z. B. eine riesige Menge von geometrischen Sätzen mitsamt deren Beweisen — fast einen Druckbogen hindurch ungefähr von Zeile zu Zeile fortschreitend — ohne jeglichen Text oder Figuren *lediglich in der Zeichensprache* dargestellt zu erblicken — nur erläutert noch durch einige ganz am Schlusse angehängte Noten nebst vorausgeschicktem Schlüssel. Die Zeichensprache ist wesentlich die unsres Klassen- und Aussagenkalkuls (mit wenigen Zufügungen), obwol äusserlich ganz eigenartig ersonnen und von der hier verfochtnen leider verschieden. Es erhellt aus ihrem Anblick, dass das S. 93 aufgestellte Ideal der Pasigraphie für die Zwecke der Wissenschaft bereits in ganz erheblichem Umfange *verwirklicht ist.* — Die Menge der sog. Axiome müsste jedenfalls noch weiter, noch sehr verringert werden. —

Peirce, Benjamin (gesprochen: Pörss).

1) Linear associative algebra, new edition with addenda and notes by Ch. S. Peirce, son of the author. New-York, Van Nostrand 1882 — Abdruck aus dem „American Journal of Mathematics", Vol. 4, p. 97 .. 229.

Peirce, Charles S(antiago).

*1) Three papers on logic, read before the American Academy of arts and sciences 1867 — siehe: „Proceedings of the American Academy of arts and sciences" Vol. 7, 1865 .. 1868:

1_a) *On an improvement in Boole's calculus of logic*, p. 250 .. 261.

1_b) On the natural classification of arguments, p. 261 .. 287.

1_c) On a new list of categories, p. 287 .. 298.

2) *Description of a notation for the logic of relatives* resulting from an amplification of the conceptions of Boole's calculus of logic. „Memoirs of the American Academy" Vol. 9, 1870, p. 317 ... 378.

3) On the application of logical analysis to multiple algebra, „Proceedings of the American Acad." 1875, Vol. 10, p. 392 .. 394.

4) Note on Grassmann's calculus of extension, ibid. 1878, Vol. 13, p. 115 sq.

*5) *On the algebra of logic,* „American Journal of Mathematics" 1880, Vol. 3, p. 15 .. 57.

6) Brief description of the algebra of relatives, 6 Seiten (wo?).

7) On the logic of number, „Amer. Journ. of Math.", Vol. 4, p. 85 .. 95.

8) *On the algebra of logic:* a contribution to the philosophy of notation. „Am. Journ. of Math." 1884, Vol. 7, p. 180 .. 202.

9) In dem Buche „Studies in logic" siehe unter S unsres Verzeichnisses:

9_a) A theory of probable inference, p. 126 .. 182.

9_b) Note A. On a limited universe of marks, p. 182 .. 187.

9_c) Note B. *The logic of relatives,* p. 187 .. 203.

*10) „Journal of speculative philosophy", Vol. 2, 1868 (three papers):

10_a) Questions concerning certain faculties claimed for man, p. 103 .. 114.

10_b) Some consequences of four incapities p. 140 .. 157.

References to Peirce

Reference to other work by Ernst Schröder.

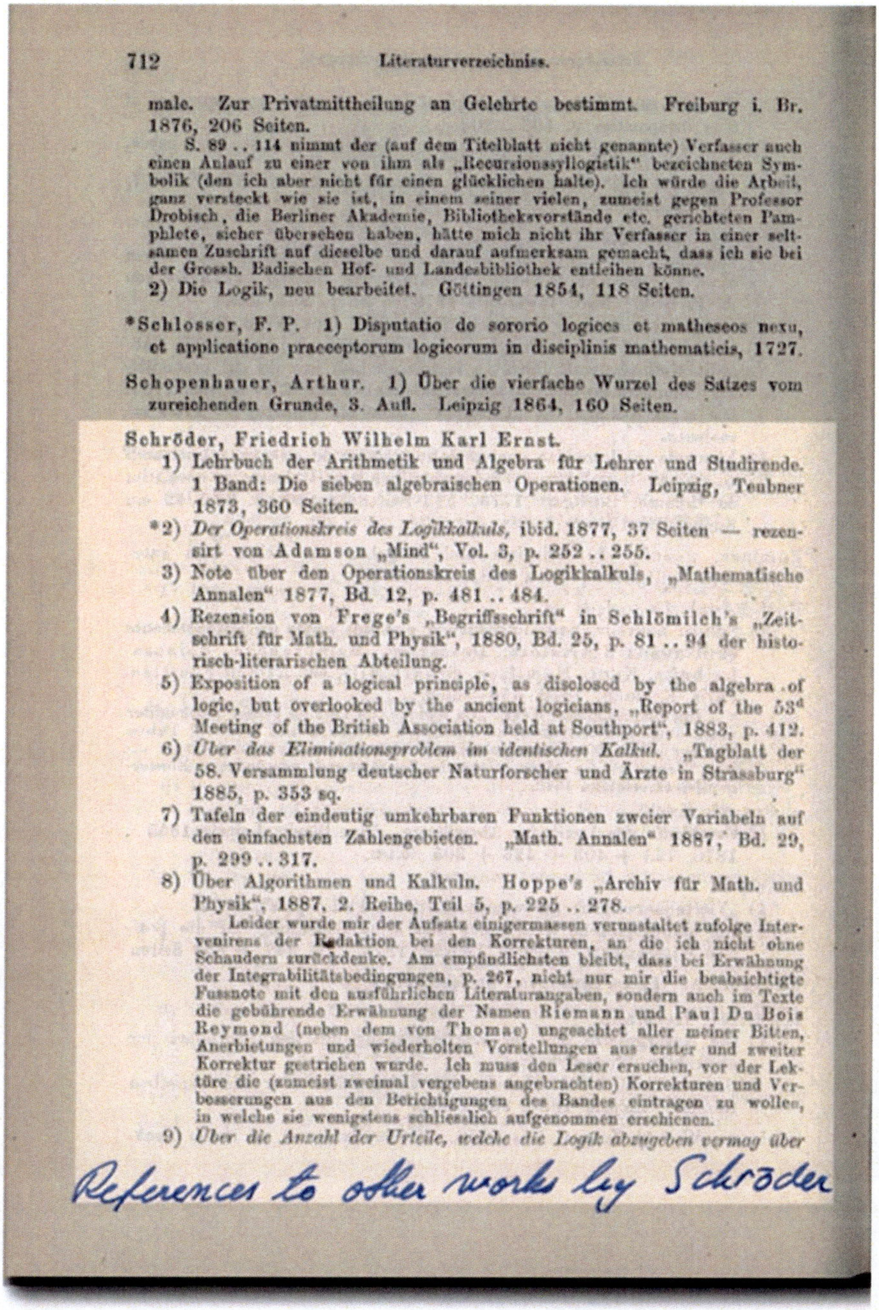

712 Literaturverzeichniss.

male. Zur Privatmittheilung an Gelehrte bestimmt. Freiburg i. Br. 1876, 206 Seiten.

S. 89 .. 114 nimmt der (auf dem Titelblatt nicht genannte) Verfasser auch einen Anlauf zu einer von ihm als „Recursionssyllogistik" bezeichneten Symbolik (den ich aber nicht für einen glücklichen halte). Ich würde die Arbeit, ganz versteckt wie sie ist, in einem seiner vielen, zumeist gegen Professor Drobisch, die Berliner Akademie, Bibliotheksvorstände etc. gerichteten Pamphlete, sicher übersehen haben, hätte mich nicht ihr Verfasser in einer seltsamen Zuschrift auf dieselbe und darauf aufmerksam gemacht, dass ich sie bei der Grossh. Badischen Hof- und Landesbibliothek entleihen könne.

2) Die Logik, neu bearbeitet. Göttingen 1854, 118 Seiten.

*Schlosser, F. P. 1) Disputatio de sororio logices et matheseos nexu, et applicatione praeceptorum logicorum in disciplinis mathematicis, 1727.

Schopenhauer, Arthur. 1) Über die vierfache Wurzel des Satzes vom zureichenden Grunde, 3. Aufl. Leipzig 1864, 160 Seiten.

Schröder, Friedrich Wilhelm Karl Ernst.

1) Lehrbuch der Arithmetik und Algebra für Lehrer und Studirende. 1 Band: Die sieben algebraischen Operationen. Leipzig, Teubner 1873, 360 Seiten.

*2) *Der Operationskreis des Logikkalkuls*, ibid. 1877, 37 Seiten — rezensirt von Adamson „Mind", Vol. 3, p. 252 .. 255.

3) Note über den Operationskreis des Logikkalkuls, „Mathematische Annalen" 1877, Bd. 12, p. 481 .. 484.

4) Rezension von Frege's „Begriffsschrift" in Schlömilch's „Zeitschrift für Math. und Physik", 1880, Bd. 25, p. 81 .. 94 der historisch-literarischen Abteilung.

5) Exposition of a logical principle, as disclosed by the algebra of logic, but overlooked by the ancient logicians, „Report of the 53[d] Meeting of the British Association held at Southport", 1883, p. 412.

6) *Über das Eliminationsproblem im identischen Kalkul.* „Tagblatt der 58. Versammlung deutscher Naturforscher und Ärzte in Strassburg" 1885, p. 353 sq.

7) Tafeln der eindeutig umkehrbaren Funktionen zweier Variabeln auf den einfachsten Zahlengebieten. „Math. Annalen" 1887, Bd. 29, p. 299 .. 317.

8) Über Algorithmen und Kalkuln. Hoppe's „Archiv für Math. und Physik", 1887, 2. Reihe, Teil 5, p. 225 .. 278.

Leider wurde mir der Aufsatz einigermassen verunstaltet zufolge Intervenirens der Redaktion bei den Korrekturen, an die ich nicht ohne Schaudern zurückdenke. Am empfindlichsten bleibt, dass bei Erwähnung der Integrabilitätsbedingungen, p. 267, nicht nur mir die beabsichtigte Fussnote mit den ausführlichen Literaturangaben, sondern auch im Texte die gebührende Erwähnung der Namen Riemann und Paul Du Bois Reymond (neben dem von Thomae) ungeachtet aller meiner Bitten, Anerbietungen und wiederholten Vorstellungen aus erster und zweiter Korrektur gestrichen wurde. Ich muss den Leser ersuchen, vor der Lektüre die (zumeist zweimal vergebens angebrachten) Korrekturen und Verbesserungen aus den Berichtigungen des Bandes eintragen zu wollen, in welche sie wenigstens schliesslich aufgenommen erschienen.

9) *Über die Anzahl der Urteile, welche die Logik abzugeben vermag über*

References to other works by Schröder

Literaturverzeichniss nebst Bemerkungen. 713

zwei Begriffe, „Tagblatt der 62. Versammlung deutscher Naturforscher und Ärzte zu Heidelberg" 1889, p. 190.

Anlässlich genannter Versammlung wurde ich erst durch Herrn Walter Dyck auf die Schriften[2] und[3] des Herrn Peano aufmerksam gemacht, nach welchen ich auch dessen Schrift[1] erwarb. Ich ersehe aus dem Vorwort der letzteren, dass die von mir ermittelte Zahl 32767 schon Herrn Peano bekannt war und in einer allgemeineren Formel desselben enthalten ist, die ich im zweiten Bande nun begründen werde.

Schuppe, Wilhelm. 1) Erkenntnisstheoretische Logik, Bonn 1878, 701 Seiten.

*Segner, J. A. 1) Specimen logicae universaliter demonstratae, 1740.

*Semler, C. A. 1) Versuch über die combinatorische Methode, 1811.

Servois. 1) Cf. Gergonne's „Annales de Mathématiques" Tome 5, p. 98, 111, 142, etc. wo sich die Namen „commutative" und „distributive" erstmalig finden.

Sigwart, Christoph.
1) *Logik*. Erster Band. Die Lehre vom Urtheil, vom Begriff und vom Schluss. Tübingen, Laupp, 1873, 420 Seiten.
2) Zweiter Band. Die Methodenlehre. Ibid. 1878, 612 Seiten.

*Solly, T. 1) Syllabus of logic, 1839.

Spalding, W. 1) Introduction to logical science, 1857.

Spottiswoode, W. 1) Remarks on some recent generalizations of algebra, „Proceedings of the London Math. society" 1872.

Steinthal, H. 1) Der Ursprung der Sprache, im Zusammenhang mit den letzten Fragen alles Wissens, eine Darstellung, Kritik und Fortentwickelung der vorzüglichsten Ansichten. 3. Ausgabe, Berlin 1877, 374 Seiten.

Stolz, Otto.
1) Vorlesungen über allgemeine Arithmetik, nach den neueren Ansichten bearbeitet. 2 Bände, Leipzig, Teubner, 1885 .. 86; 344 + 326 Seiten.

Studies in logic by members of the Johns Hopkins University. Boston, Little, Brown & Co., 1883, 203 Seiten.

Sweet, Henry. 1) Words, logic and grammar, „Transactions of Philological society", 1876.

Thomson, W. 1) Laws of thought, 1875.

Thoughts on logic, or the S. N. I. X. propositional theory, 1877.

*Tönnies. 1) De logicae scientiae ad exemplum arithmeticae instituenda ratione, 1752.

Trendelenburg, Adolf.
1) Logische Untersuchungen, 2 Bde, Leipzig 1870, 3. Aufl. 388 + 538 Seiten.

Reference to John Venn and several other authors.

714 Literaturverzeichniss.

2) Historische Beiträge zur Philosophie, 4 Bde. Band 3, Berlin 1867, 444 Seiten.

Twesten, A. D. C. 1) Logik 1825.

Ueberweg, Friedrich.
1) System der Logik und Geschichte der logischen Lehren. 4. Aufl., Bonn, Marcus 1874, 434 Seiten.

Ulrich, J. H. 1) Institutiones logicae et metaphysicae, 1792.

Ulrici, H. „Zeitschrift für Philosophie und philosophische Kritik", 1878.

Venn, John.
1) *Symbolic Logic,* London, Macmillan, 1881; 446 Seiten.
Wegen der ausserordentlichen Belesenheit des Verfassers, seiner sorgfältigen kritischen Anmerkungen und seiner „Historic notes" in Chapter XX, in Bezug auf die Entwickelungsgeschichte der symbolisirenden Logik eine schätzenswerte Ergänzung zum vorliegenden Buche.
2) The logic of chance, an essay on the foundations and province of the theory of probability with especial reference to its logical bearings and its application to moral and social science. 2[d] ed. London, Macmillan, 1876, 488 Seiten.
3) Consistency and real inference, „Mind", Vol. 1, 1876, p. 43 .. 52.
*4) Boole's logical system, ibidem p. 479 .. 491.
*5) On the diagrammatic and mechanical representation of propositions and reasonings (The London, Edinbourgh and Dublin), „Philosophical Magazine" (and Journal of Science), Vol. 10, 5th series, 1880, p. 1..18.
*6) Symbolic logic, „Princeton Review", New York, Sept. 1880, p. 247..267.
*7) On the various notations adopted for expressing the common propositions of logic, „Proceedings of the Cambridge Philosophical society", Dec. 1880, Vol. 4, p. 35 .. 46.
*8) On the employment of geometrical diagrams for the sensible representation of logical propositions, ibid. p. 46 .. 58.
9) The difficulties of material logic, „Mind" Vol. 4, p. 35 .. 47.
10) On the forms of logical proposition, „Mind" Vol. 5, p. 336 .. 349.
In 4) bis 8) sind einzelne Kapitel von 1) vorausbearbeitet.
11) The principles of empirical or inductive logic, Macmillan 1889, 594 Seiten.
Enthält auch viel zur formalen Logik gehöriges, u. a. schätzenswerte Angaben über Universalsprachen.

Vives, Ludwig. 1) De censura veri, 1555.

Voigt, Andreas Heinrich.
1) Die Auflösung von Urteilssystemen, das Eliminationsproblem und die Kriterien des Widerspruchs in der Algebra der Logik. Freiburger Doktordissertation, Leipzig, Alex. Danz 1890.

Waitz, Theodor. 1) Lehrbuch der Psychologie als Naturwissenschaft, Braunschweig 1849, 685 Seiten.

Weber, Heinrich. 1) Über Causalität in den Naturwissenschaften.

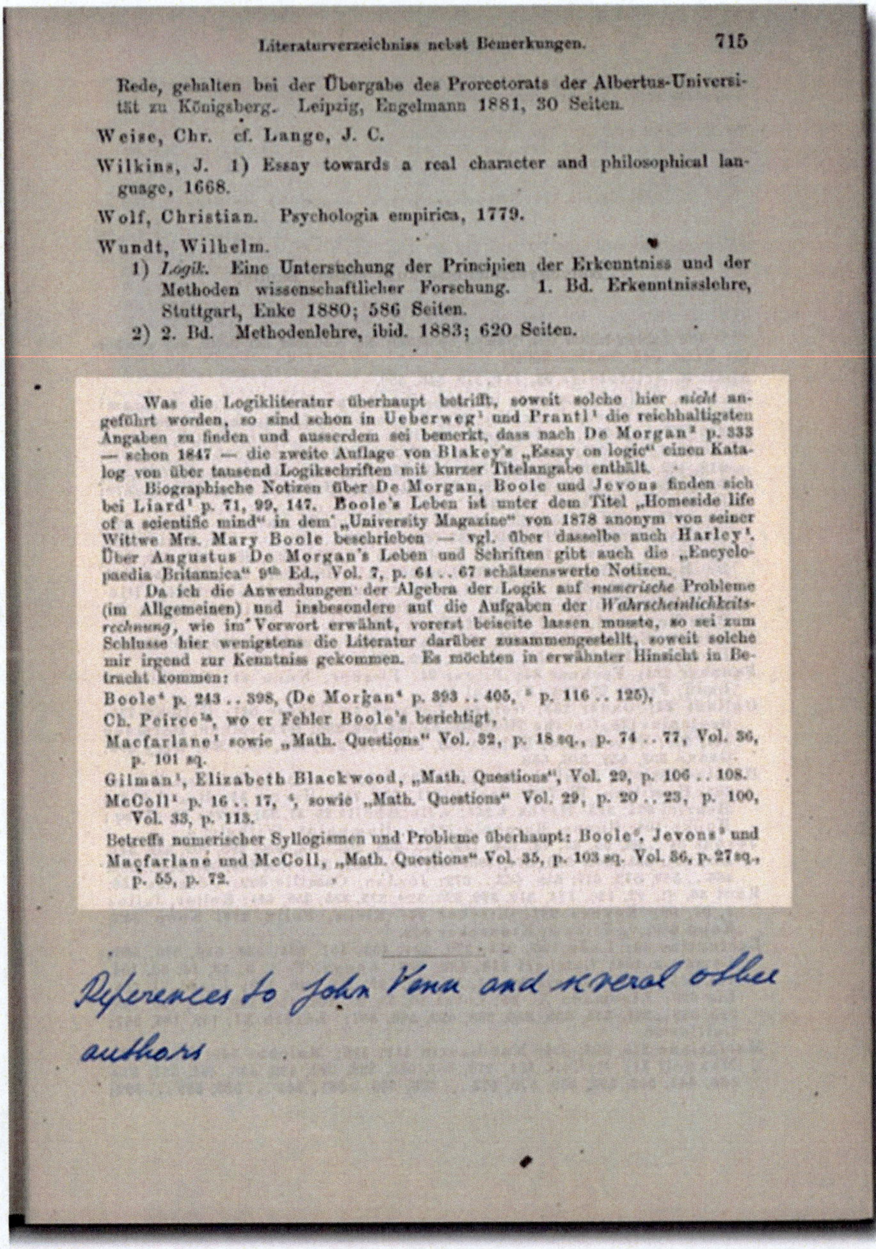

Literaturverzeichniss nebst Bemerkungen. 715

Rede, gehalten bei der Übergabe des Prorectorats der Albertus-Universität zu Königsberg. Leipzig, Engelmann 1881, 30 Seiten.

Weise, Chr. cf. Lange, J. C.

Wilkins, J. 1) Essay towards a real character and philosophical language, 1668.

Wolf, Christian. Psychologia empirica, 1779.

Wundt, Wilhelm.
1) *Logik.* Eine Untersuchung der Principien der Erkenntniss und der Methoden wissenschaftlicher Forschung. 1. Bd. Erkenntnisslehre, Stuttgart, Enke 1880; 586 Seiten.
2) 2. Bd. Methodenlehre, ibid. 1883; 620 Seiten.

Was die Logikliteratur überhaupt betrifft, soweit solche hier *nicht* angeführt worden, so sind schon in Ueberweg[1] und Prantl[1] die reichhaltigsten Angaben zu finden und ausserdem sei bemerkt, dass nach De Morgan[2] p. 333 — schon 1847 — die zweite Auflage von Blakey's „Essay on logic" einen Katalog von über tausend Logikschriften mit kurzer Titelangabe enthält.

Biographische Notizen über De Morgan, Boole und Jevons finden sich bei Liard[1] p. 71, 99, 147. Boole's Leben ist unter dem Titel „Homeside life of a scientific mind" in dem „University Magazine" von 1878 anonym von seiner Wittwe Mrs. Mary Boole beschrieben — vgl. über dasselbe auch Harley[1]. Über Augustus De Morgan's Leben und Schriften gibt auch die „Encyclopaedia Britannica" 9th Ed., Vol. 7, p. 64 .. 67 schätzenswerte Notizen.

Da ich die Anwendungen der Algebra der Logik auf *numerische* Probleme (im Allgemeinen) und insbesondere auf die Aufgaben der *Wahrscheinlichkeitsrechnung*, wie im Vorwort erwähnt, vorerst beiseite lassen musste, so sei zum Schlusse hier wenigstens die Literatur darüber zusammengestellt, soweit solche mir irgend zur Kenntniss gekommen. Es möchten in erwähnter Hinsicht in Betracht kommen:

Boole[4] p. 243 .. 398, (De Morgan[4] p. 393 .. 405, [5] p. 116 .. 125),
Ch. Peirce[18], wo er Fehler Boole's berichtigt,
Macfarlane[1] sowie „Math. Questions" Vol. 32, p. 18 sq., p. 74 .. 77, Vol. 36, p. 101 sq.
Gilman[1], Elizabeth Blackwood, „Math. Questions", Vol. 29, p. 106 .. 108.
McColl[1] p. 16 .. 17, [4], sowie „Math. Questions" Vol. 29, p. 20 .. 23, p. 100, Vol. 33, p. 113.
Betreffs numerischer Syllogismen und Probleme überhaupt: Boole[6], Jevons[3] und Macfarlane und McColl, „Math. Questions" Vol. 35, p. 103 sq. Vol. 36, p. 27 sq., p. 55, p. 72.

References to John Venn and several other authors

Charles Sanders Peirce

3.2 Contributions by Ch. S. Peirce

Peirce, Ch.S., "Description of a notation for the logic of relatives, resulting from an amplification of the conceptions of Boole's calculus of logic", *Memoirs of the American Academy*, 9, 1870, 317-378.

Peirce, Ch.S., "On the algebra of logic", *American Journal of Mathematics*, Vol. 3, No. 1, March 1880, 15-57.

Peirce, Ch.S., "On the algebra of logic: A Contribution to the philosophy of notation", *American Journal of Mathematics*, Vol. 7, No. 2, January 1885, 180-196.

Peirce, Ch.S., "On the algebra of logic [Continued]", *American Journal of Mathematics*, Vol. 7, No. 3, April 1885, 197-202.

Peirce arrow = Logic NOR

Carolyn Eisele, "Mathematical miscellanea by Charles S. Pierce", *The New Elements of Mathematics*, Vol. 3, Parts 1 and 2, Mouton Publishers, The Hague and Paris, and Humanities Press, Atlantic Highlands, N.J., 1976, xxxix + 1153 pp.

Carolyn Eisele, "Mathematical philosophy by Charles S. Pierce", *The New Elements of Mathematics*, Volume 4, Mouton Publishers, The Hague and Paris, and Humanities Press, Atlantic Highlands, N.J., 1976, xxviii + 393 pp.
Reviewed by Jay Zeman, *The Journal of Symbolic Logic*, Vol. 47, No. 3, September 1982, 705-708.

The *Appolo Guidance Computer* used in the spaceship that first carried humans to the moon, was constructed entirely using NOR gates with three inputs.

On the Algebra of Logic.

By C. S. Peirce.

Chapter I. — Syllogistic.

§ 1. *Derivation of Logic.*

In order to gain a clear understanding of the origin of the various signs used in logical algebra and the reasons of the fundamental formulæ, we ought to begin by considering how logic itself arises.

Thinking, as cerebration, is no doubt subject to the general laws of nervous action.

When a group of nerves are stimulated, the ganglions with which the group is most intimately connected on the whole are thrown into an active state, which in turn usually occasions movements of the body. The stimulation continuing, the irritation spreads from ganglion to ganglion (usually increasing meantime). Soon, too, the parts first excited begin to show fatigue; and thus for a double reason the bodily activity is of a changing kind. When the stimulus is withdrawn, the excitement quickly subsides.

It results from these facts that when a nerve is affected, the reflex action, if it is not at first of the sort to remove the irritation, will change its character again and again until the irritation is removed; and then the action will cease.

Now, all vital processes tend to become easier on repetition. Along whatever path a nervous discharge has once taken place, in that path a new discharge is the more likely to take place.

Accordingly, when an irritation of the nerves is repeated, all the various actions which have taken place on previous similar occasions are the more likely to take place now, and those are most likely to take place which have most frequently taken place on those previous occasions. Now, the various actions which did not remove the irritation may have previously sometimes been performed and sometimes not; but the action which removes the irritation must

First page of the paper
Peirce, Ch., S., "On the algebra of logic", *American Journal of Mathematics*, Vol. 3, 1880, 15-57.

On the Algebra of Logic:

A CONTRIBUTION TO THE PHILOSOPHY OF NOTATION.

BY C. S. PEIRCE.

I.—*Three kinds of Signs.*

Any character or proposition either concerns one subject, two subjects, or a plurality of subjects. For example, one particle has mass, two particles attract one another, a particle revolves about the line joining two others. A fact concerning two subjects is a dual character or relation; but a relation which is a mere combination of two independent facts concerning the two subjects may be called *degenerate*, just as two lines are called a degenerate conic. In like manner a plural character or conjoint relation is to be called degenerate if it is a mere compound of dual characters.

A sign is in a conjoint relation to the thing denoted and to the mind. If this triple relation is not of a degenerate species, the sign is related to its object only in consequence of a mental association, and depends upon a habit. Such signs are always abstract and general, because habits are general rules to which the organism has become subjected. They are, for the most part, conventional or arbitrary. They include all general words, the main body of speech, and any mode of conveying a judgment. For the sake of brevity I will call them *tokens*.

But if the triple relation between the sign, its object, and the mind, is degenerate, then of the three pairs sign object
sign mind
object mind
two at least are in dual relations which constitute the triple relation. One of the connected pairs must consist of the sign and its object, for if the sign were not related to its object except by the mind thinking of them separately, it would not fulfil the function of a sign at all. Supposing, then, the relation of the sign to its object does not lie in a mental association, there must be a direct dual

First page of the paper Peirce, Ch. S., "On the algebra of logic - A contribution to the philosophy of notation", *American Journal of Mathematics*, Vol. 7, 1884, 180-203.

John Venn

3.3 Venn Diagrams

Venn, J., "On the diagrammatic and mechanical representation of propositions and reasonings", *Dublin Philosophical Magazine and Journal of Science*, Vol. 9, No. 59, 1880, 1-18.

Venn, J., *Symbolic Logic*, Macmillan, London, 1881, 446 pages.

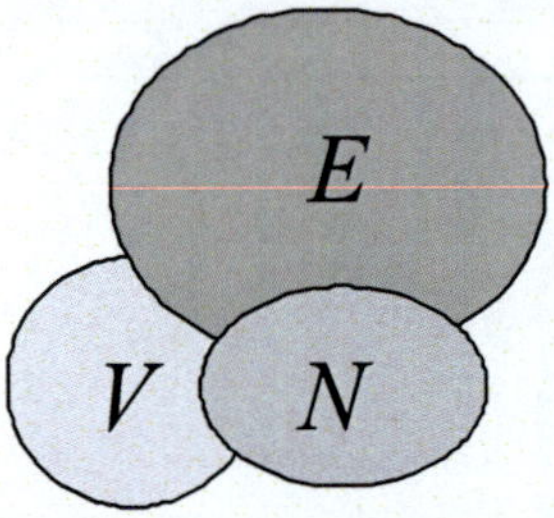

Venn introduced a way for diagramming notation by Boole, which is now called the *Venn diagrams,* and used to express all possible relationships between sets (collections of objects).

A painting of John Venn by Charles E. Brock.
Photograph by Christopher Hurst, Hamilton -Kerr Institute, University of Cambridge.

Visualization of logical relationships by Venn, have been elaborated by Charles Lutwidge Dodgson, more widely known as Carroll Lewis, the author of famous children novels, including for example *Alice in Wonderland*and *Through the Looking Glass*

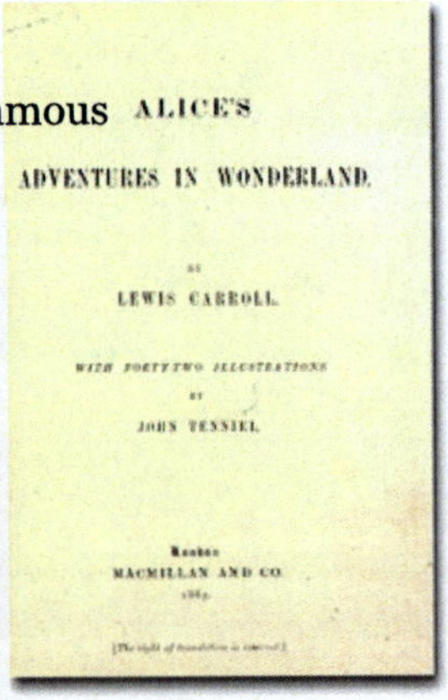
ALICE'S

ADVENTURES IN WONDERLAND.

LEWIS CARROLL.

JOHN TENNIEL.

MACMILLAN AND CO.

Chapter 4
Boolean Algebra – A Mathematical Subject

R.S. Stanković & J. Astola: From Boolean Log. to Switch. Circuits & Automata, SCI 335, pp. 111–117.
springerlink.com

Edward Vermilye Huntington

4.1 Axiomatization by Huntington

Huntington, E.V., "Sets of Independent Postulates for the Algebra of Logic", *Transactions of the American Mathematical Society*, Vol. 5, No. 3, July 1904, 288-309.

Boolean algebra presented as an axiomatic algebraic structure, in 1904.

SETS OF INDEPENDENT POSTULATES FOR THE ALGEBRA OF LOGIC*

BY

EDWARD V. HUNTINGTON

The algebra of symbolic logic, as developed by LEIBNIZ, BOOLE, C. S. PEIRCE, E. SCHRÖDER, and others,† is described by WHITEHEAD as "the only known member of the non-numerical genus of universal algebra."‡ This algebra, although originally studied merely as a means of handling certain problems in the logic of classes and the logic of propositions, has recently assumed some importance as an independent calculus; it may therefore be not without interest to consider it from a purely mathematical or abstract point of view, and to show how the whole algebra, in its abstract form, may be developed from a selected set of fundamental propositions, or postulates, which shall be independent of each other, and from which all the other propositions of the algebra can be deduced by purely formal processes.

In other words, we are to consider the construction of a purely *deductive theory*, without regard to its possible applications.

Introductory remarks on deductive theories in general.§ The first step in such a discussion is to decide on the *fundamental concepts* or *undefined symbols*, concerning which the statements of the algebra are to be made.

One such concept, common to every mathematical theory, is the notion of

1) a *class* (K) of *elements* $(a, b, c, \cdots)$. ‖

*Presented to the Society: §1, September 1, 1903; §2, December 28, 1903 (and since revised); §3 and the appendix, April 30, 1904. Received for publication, April 30, 1904.

† For an extensive bibliography, see SCHRÖDER's *Algebra der Logik*, vol. 1 (1890).

‡ A. N. WHITEHEAD, *Universal Algebra*, vol. 1 (1898), p. 35.

§ Cf. papers by A. PADOA, cited in Transactions, vol. 4 (1903), p. 358.

‖ A *class* is determined by stating some condition which every entity in the universe must either satisfy or not satisfy; every entity which satisfies the condition is said to *belong* to the class. (If the condition is such that no entity can satisfy it, the class is called a "null" class.) Every entity which belongs to the class in question is called an *element* (cf. H. WEBER, *Algebra*, vol. 2 (1899), p. 3).

No further analysis of this concept *class*, or of similar concepts introduced below, is here attempted. For an elaborate discussion of the logical processes which underlie all mathematical thinking, see B. RUSSELL's work on *The Principles of Mathematics*, vol. 1, 1903.

BOOLEAN ALGEBRA. A CORRECTION

BY

EDWARD V. HUNTINGTON

In my paper in these Transactions for January, 1933, the Example 4.5 on page 286 is erroneous, and Postulate 4.5 on page 280 is in fact redundant. Hence the "fourth set" of postulates for Boolean algebra, on the base $(K, +, ')$, should read as follows (the class K being understood to contain at least two distinct elements):

POSTULATE 4.1. *If a and b are in K, then $a+b$ is in K.*

POSTULATE 4.2. *If a is in K, then a' is in K.*

POSTULATE 4.3. $a+b=b+a$.

POSTULATE 4.4. $(a+b)+c=a+(b+c)$.

POSTULATE 4.6. $(a'+b')'+(a'+b)'=a$ [or, $ab+ab'=a$, where, by definition, $ab=(a'+b')'$].

The steps by which the proposition 4.5 ($a+a=a$) is deduced as a theorem from Postulates 4.1, 4.2, 4.3, 4.4, and 4.6, are as follows.*

4.10. $a''=a$. (Proof as on page 281.)

4.11. $a+a'=b+b'$.

Proof (without using 4.5). By 4.6, with 4.3 and 4.4,

$$a + a' = [(a' + b'')' + (a' + b')'] + [(a'' + b'')' + (a'' + b')']$$
$$= [(b' + a'')' + (b' + a')'] + [(b'' + a'')' + (b'' + a')'] = b + b'.$$

4.12. Definition. $U=a+a'=$ the "universe element" of the system.

In particular, $U=U+U'$.

4.15. $a+U'=a$.

By 4.6, 4.10, 4.12,

(a) $$U' = (U + U)' + (U + U')' = (U + U)' + U'.$$

By 4.12, (a), 4.4, 4.12,

$$U = U + U' = U + [(U + U)' + U']'$$
$$= (U + U') + (U + U)' = U + (U + U)'.$$

Hence by 4.4, 4.12,

(b) $$U + U = U + [U + (U + U)'] = (U + U) + (U + U)' = U.$$

From (a), (b), $U'=U'+U'$, whence by 4.12,

* For an essential step in this proof I am indebted to Mr. B. Notcutt, a Commonwealth Fellow from Oxford University, at present a graduate student in Harvard University. The following article appeared after my brief bibliography was completed: W. V. Quine, *A note on Nicod's postulate*, Mind, vol. 41 (1932), pp. 345–350.

557

Huntington, E.V., "Boolean Algebra. A Correction", *Transactions of the American Mathematical Society*, Vol. 35, No. 2, April 1933, 557-558.

A SECOND CORRECTION

BY

EDWARD V. HUNTINGTON

In my paper in the present volume these Transactions (vol. 35, pp. 274–304 and 557–558), the Example 5 on page 304 is erroneous and Postulate 5 on page 301 is redundant. That is, Postulates 1, 2, 3, 4, 6, 7 (without 5) form a set of independent postulates for the "informal" system of *Principia Mathematica*.

The proof of 5 from 1, 2, 3, 4, 6, 7 is as follows.*

6a. *If $a+b$ is in T and a not in T, then b is in T.* (From 6.)

6b. *If a not in T and b not in T, then $a+b$ not in T.* (From 6.)

7a. *If a is in T, then a' is not in T.* (From 7.)

3a. *If b is in T, then $a+b$ is in T.*

For, by 7a, b' is not in T. But by 3, $b'+(a+b)$ is in T. Hence by 6a, $a+b$ is in T.

4a. *If b is in T, then $b+a$ is in T.*

For, by 3a, $a+b$ is in T, whence by 7a, $(a+b)'$ is not in T. But by 4, $(a+b)'+(b+a)$ is in T. Hence by 6a, $b+a$ is in T.

5a. *If a is not in T, then a' is in T.*

For, suppose a' not in T. Then by 6b, $a'+a$ not in T, whence by 6b, $a'+(a'+a)$ not in T, contrary to 3.

5. *If a, b, etc. are in K, then $(b'+c)'+[(a+b)'+(a+c)]$ is in T.*

Case 1; *a in T.* By 4a, $a+c$ is in T. Hence the theorem, by 3a (twice).

Case 2: *b in T.* By 7a, b' is not in T. If c is in T, then by 3a, $a+c$ is in T, whence the theorem, by 3a (twice). If c is not in T, then by 6b, $b'+c$ is not in T, whence by 5a, $(b'+c)'$ is in T, whence the theorem, by 4a.

Case 3: *a not in T and b not in T.* By 6b, $a+b$ not in T, whence by 5a, $(a+b)'$ is in T. Hence the theorem, by 4a and 3a.

The proof is thus complete. It can also be shown that 1, 2, 3a, 4a, 5a, 6a, 7 form a set of independent postulates equivalent to the set 1, 2, 3, 4, 6, 7.

---

* For valuable suggestions in this connection I am indebted to Professor Alonzo Church and Dr. K. E. Rosinger.

HARVARD UNIVERSITY,
CAMBRIDGE, MASS.
May 29, 1933.

Huntington, E.V., "A Second Correction",
Transactions of the American Mathematical Society,
Vol. 35, No. 4, October 1933, 971.

4.2 Boolean Algebra – A Strong Mathematical Subject

Stone, Birkhoff, MacLane

Marshall Harvey Stone in 1936, Garrett Birkhoff in 1940, and Birkhoff and MacLane in 1941, raised the Boolean algebra to the level of deep mathematical subjects.

Stone, M.H., "The theory of representation for Boolean algebras",*Transactions of the American Mathematical Society*, Vol. 40, 1936, 37-111.

Stone, M.H., "The representation of Boolean algebras", *Bulletin of the American Mathematical Society*, Vol. 44, 1938, 807-816.

M. H. Stone did his PhD dissretation under supervision of George David Birkhoff the father of Garrett Birkhoff.

Birkhoff, G., "Lattice theory",*Amer. Math. Soc.*, Colloquium Publications, Vol. 25, 1940, Chapters 5 and 6.

Birkhoff, G., Mac Lane, S., *A Survey of Modern Algebra*, New York, Mcmillan Co., 1941.

Saunders Mac Lane wrote a PhD Thesis in Mathematical Logic at the Mathematical Institute of Gottingen in 1934.

Chapter 5
Application of Algebra of Logic in Engineering

R.S. Stanković & J. Astola: From Boolean Log. to Switch. Circuits & Automata, SCI 335, pp. 121–128.
springerlink.com

Paul S. Ehrenfest

5.1 Remarks and Ideas by P. Ehrnfest

The application of Boolean algebra to the analysis of electric-relay contact circuits was first noted by the Russian physicist P.S. Ehrenfest in 1910 in a review of the work by Louis Couturat

Ehrenfest, P., "Review of Couturat's Algebra logiki", Žurnal Russkago Fiziko-hemičeskago Obščestva, Fizičeskij otdel, Vol. 42, 1910,Otdel vtoroj, 382-387.

Couturat, L.,*L'algebre de la logique*, Paris 1905, Volume number 24 in Gauthier-Villars collection *Scientia*, 100 pages, 2nd. edn., Paris 1914, 100 pages.

Hungarian translation *A logika algebraja*, translated by Denes Konig, Mathematikai es physikai lapok, Budapest, Vol. 17, 1908, 109-202,

Russian translation *Algebra logiki*, Mathesis, Odessa, 1909, iv+107+xii+6.

English version *The Algebra of Logic*, translated by Lydia G. Robinson and Philip E. B. Jourdain, The Open Court Publishing Company, Chikago, 1914, xiv + 98 pages, price $1.50

Reviewed by James Byrnie in *Amer. Math. Monthly*, Vol. 22, No. 3, March 1915, 95-97.

Claro C. Dassen,"Vida y Obra de Luis Couturat", *Anales de la Academia National de Ciencias Exactas, Fisicas, y Natruales de Buenos Aires*, Vol. 4, 1939, 73-204.

It is written in the review of this paper by W. VQuine, *The Journal of Symbolic Logic*, Vol. 5, No. 4, December 1940, 168-169

Logicians know Couturat as an expositor and defender of a new logic, as an commentator of the philosophy of mathematics, and as a pioneer editor and analyst of the logic of Leibniz.

This statement about Ehrenfest has been pointed out in [Zinoviev, 1957], and [Zinoviev, 1958], where it is reported that in a paper presented at the Seminar on Logic at the Institute of Logic of the Academy of Science of SSSR, Geleii Nikolaevic Povarov discussed combinatory logic, distinguishing *technical logic* from the *algebra of logic*and pointing out that *the possibility of technical applications of mathematical logic* (i.e., the application of Boolean algebra to the analysis of electric relay-contact circuits) was first noted by the Russian physicist P.S. Ehrenfest in 1910 in a review of the work by Louis Couturat. See also Comey, Kline. The same point of view has been supported also by Durand-Richard.

Zinoviev, A.A., "Rasšcirat tematiku logiceskih issledovanij (Broaden the subject matter of logical investigations)", *Voprosy filosofii*, No. 3, 1957, 211-215.

Zinoviev, A.A., "O rabote seminara po logike v Institute FilosofiiAN SSSR (Work of the Seminar of Logic in the Institute of Philosophy of the Academy of Sciences of the SSSR)", *Voprosy filosofii*, No. 2, 1958, 162-172.

Comey, D.D., Kline, G.L., "Review of *Rabote seminara po logiki v Institute Filosofii AN SSSR* (Work of the Seminar on Logic in the Institute of Philosophy of the Academy of Sciences of the USSSR) by A. A. Zinoviev", *The Journal of Symbolic Logic*, Vol. 24, No. 3, 1959, 232-233.

Durand-Richard, M.J., "Pour une historiographie comparée de l'histoire de l'informatique", Report, Université Paris 8-REHSEIS.

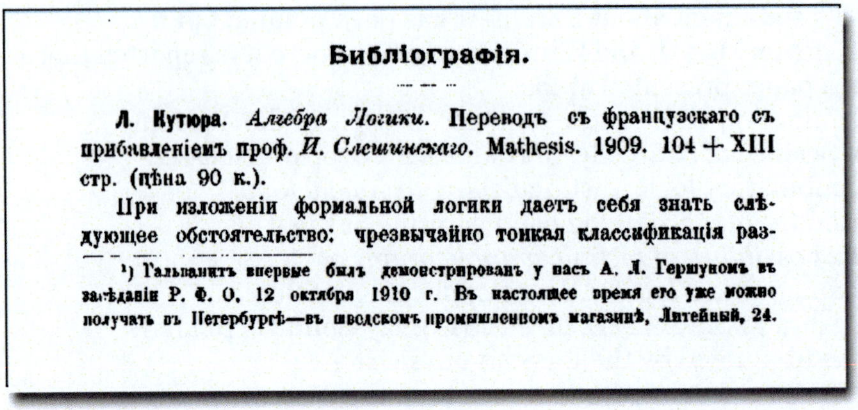

Библіографія.

Л. Кутюра. *Алгебра Логики.* Переводъ съ французскаго съ прибавленіемъ проф. *И. Слешинскаго.* Mathesis. 1909. 104 + XIII стр. (цѣна 90 к.).

При изложеніи формальной логики даетъ себя знать слѣдующее обстоятельство: чрезвычайно тонкая классификація раз-

[1]) Гальванитъ впервые былъ демонстрированъ у насъ А. Л. Гершуномъ въ засѣданіи Р. Ф. О. 12 октября 1910 г. Въ настоящее время его уже можно получать въ Петербургѣ—въ шведскомъ промышленномъ магазинѣ, Литейный, 24.

Pages 382 of the review by Ehrnfest.

An English translation of the complete
Review by Paul Ehrnfest of the book by Couturat
will appear in TICSP Series, 2009.

— 383 —

личныхъ типовъ сужденій и умозаключеній, выработанная уже въ *сознаніи*, находитъ въ *языкѣ* слишкомъ тяжеловѣсный и неточный инструментъ для своего выраженія.

Поэтому въ ученіи о сужденіяхъ и умозаключеніяхъ уже давно принято передавать эту классификацію условными символами.

Сюда относятся прежде всего подраздѣленіе сужденій на типы „*А, Е, И, О*" и произведенные изъ нихъ словесные символы для 19 формъ правильныхъ умозаключеній (отъ „Барбара" до „Феризонъ" съ 13-го столѣтія). Позднѣе развилась символика, представляющая различныя понятія при помощи круговъ на плоскости: различные способы относительнаго расположенія круговъ соотвѣтствуютъ различнымъ случаямъ соединенія двухъ понятій въ одно сужденіе.

Первая символика едва ли болѣе, чѣмъ стенографія, объединяющая въ одну картину всѣ члены сужденія, но негибкая. Вторая символика даетъ уже гораздо большее: надъ системой такихъ круговъ можно по опредѣленнымъ правиламъ производить преобразованія, имѣющія также опредѣленную интерпретацію въ логикѣ. (Сравнить съ этимъ тотъ фактъ, что формулы химіи доставляютъ не только систематическую регистратуру различныхъ веществъ, но что кромѣ этого—преобразованія этихъ формулъ, произведенныя по опредѣленнымъ правиламъ, соотвѣтствуютъ химическимъ преобразованіямъ).

Само собою разумѣется, что эта область рано или поздно должна была побудить *спекулятивно*-математическіе умы къ слѣдующей попыткѣ: тотъ принципъ символическихъ обозначеній, который оказался столь плодотворнымъ для *дѣйствій надъ числами и величинами*—„буквенное исчисленіе" — перенести на *дѣйствія надъ всякими понятіями*.

Сюда относятся уже попытки Лейбница и Грассмана. Болѣе широкимъ развитіемъ обязана „*алгебра логики*" двумъ математикамъ: англичанину Джорджу Булю (1815—1864) и нѣмцу Е. Шредеру (1841—1902); въ дальнѣйшей разработкѣ принимали участіе математики всѣхъ странъ. Въ частности русскій математикъ П. Порѣцкій много содѣйствовалъ упрощенію методовъ своей оригинальной постановкой вопроса.

„*Алгебра логики*" прежде всего устанавливаетъ символическія обозначенія для тѣхъ элементарныхъ дѣйствій, которыя являются такими же основными при оперированіи съ понятіями, какъ сло-

— 387 —

основывается формальная логика, выступаютъ въ отчетливой и притомъ совершенно непривычной формѣ.

2. Вслѣдствіе того, что формулировка всякаго разсужденія при помощи логическихъ равенствъ по крайней мѣрѣ въ 5—10 разъ короче словесной, достигается удивительная сжатость изложенія.

3. Символическая формулировка даетъ возможность «вычислять» слѣдствія изъ такихъ сложныхъ системъ посылокъ, въ которыхъ при словесномъ изложеніи почти или совершенно невозможно разобраться.

Къ счастью, уже отвыкли требовать отъ каждой математической спекуляціи прежде всего «практической пользы». Тѣмъ не менѣе, быть можетъ, умѣстно коснуться вопроса о томъ, не встрѣчаются ли въ физикѣ или въ техникѣ въ самомъ дѣлѣ такія сложныя системы посылокъ. *Мнѣ думается, что на этотъ вопросъ слѣдуетъ отвѣтить утвердительно.* Примѣръ: пусть имѣется проектъ схемы проводовъ автоматической телефонной станціи. Нужно опредѣлить: 1) будетъ ли она правильно функціонировать при любой комбинаціи, могущей встрѣтиться въ ходѣ дѣятельности станціи; 2) не содержитъ ли она излишнихъ усложненій.

Каждая такая комбинація является посылкой, каждый маленькій коммутаторъ есть логическое «или—или», воплощенное въ эбонитѣ и латуни; все вмѣстѣ—система чисто качественныхъ (въ сѣти слабаго тока именно *не* количественныхъ) «посылокъ», ничего не оставляющая желать въ отношеніи сложности и запутанности.

Слѣдуетъ ли при рѣшеніи этихъ вопросовъ разъ навсегда удовлетвориться геніальнымъ—а по большей части просто рутиннымъ—способомъ *пробованія на графикѣ*?

Правда ли, что несмотря на существованіе уже разработанной «алгебры логики» своего рода «алгебра распредѣлительныхъ схемъ» должна считаться утопіей?

П. Эренфестъ.

А. В. Цингеръ. *Начальная физика.* Первая ступень. XX+499 стр. Москва.

„Первая ступень" курса физики, разсчитаннаго авторомъ на двѣ части, представляетъ собою попытку „дать элементарный очеркъ физики, имѣющій цѣлью выяснить начинающимъ немногіе

4

A wonderful comment – Is it right to consider the specific algebra of Schw Switching networks as a utopia, in spite of the already developed algebra of logic

Chapter 6
Switching Theory – From Art and Skills to Scientific Methods

R.S. Stanković & J. Astola: From Boolean Log. to Switch. Circuits & Automata, SCI 335, pp. 129–140.
springerlink.com

Claude Elwood Shannon

6.1 Switching Theory from Boolean Algebra

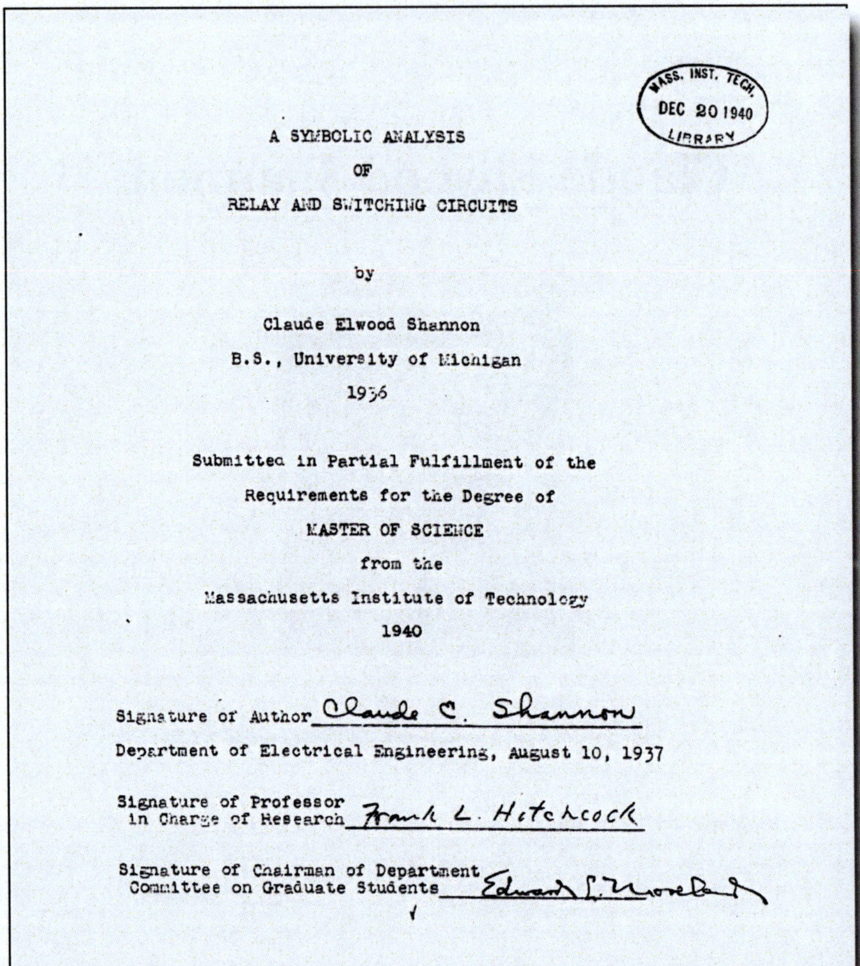

A SYMBOLIC ANALYSIS
OF
RELAY AND SWITCHING CIRCUITS

by

Claude Elwood Shannon
B.S., University of Michigan
1936

Submitted in Partial Fulfillment of the
Requirements for the Degree of
MASTER OF SCIENCE
from the
Massachusetts Institute of Technology
1940

Signature of Author Claude E. Shannon
Department of Electrical Engineering, August 10, 1937

Signature of Professor
in Charge of Research Frank L. Hitchcock

Signature of Chairman of Department
Committee on Graduate Students

The first page of the MSC Thesis by C.E. Shannon that converted logic circuit design from arts and skills into a scientific discipline by assuming the Boolean algebra as the underlying algebraic structure to represent switching (Boolean) functions.

38

TABLE OF CONTENTS

Page

I Introduction: Types of Problems - - - - - - - - - - 1

II Series-Parallel Two-Terminal Circuits - - - - - - 4

Fundamental Definitions and Postulates - - - - - - 4

Theorems - 6

Analogue with the Calculus of Propositions - - - - - 8

III Multi-Terminal and Non-Series-Parallel Networks - - -18

Equivalence of n-Terminal Networks - - - - - - - - - 18

Star-Mesh and Delta-Wye Transformations - - - - - - 19

Hinderance Function of a Non-Series-Parallel Network 21

Simultaneous Equations - - - - - - - - - - - - - - 24

Matrix Methods - - - - - - - - - - - - - - - - - - - 25

Special Types of Relays and Switches - - - - - - - - 28

IV Synthesis of Networks - - - - - - - - - - - - - - - 31

General Theorems on Networks and Functions - - - - 31

Dual Networks - - - - - - - - - - - - - - - - - - - 36

Synthesis of the General Symmetric Function - - - - 39

Equations from Given Operating Characteristics - - - 47

V Illustrative Examples - - - - - - - - - - - - - - - -51

A Selective Circuit - - - - - - - - - - - - - - - - - 52

An Electric Combination Lock - - - - - - - - - - - 55

A Vote Counting Circuit - - - - - - - - - - - - - - 58

An Adder to the Base Two - - - - - - - - - - - - - 59

A Factor Table Machine - - - - - - - - - - - - - - 62

References - 69

240579

Contents of the MSc Thesis of Shannon defended at MIT in 1938.
By many scholars this is the most referenced Msc thesis in the last century.

Majority of scholars working towards mathematical foundations for analysis and design of relay circuits focussed their efforts towards study of many particular examples attempting to disclose mathematical relationships describing such circuits. Many of them realized latter that partial mathematical results which they have developed are identical to elements of the Boolean algebra.

Unlike to this, Shannon realized the analogue between the symbolic analysis of realy circuits and calculus of propositions and due to that exploited the Boolean algebra as mathematical foundations for realy circuit analysis and design.

These results were first reported in his MSc Thesis in 1938

[Ref. Shannon 1]
and latter published in

Shannon, C.E., "A symbolic analysis of relay and switching circuits", *Transactions of American Institute of Electrical Engineers*, Vol. 57, 1938.

In this paper, Shannon used the following Table of postulates of the Boolean algebra [Ref. Shannon 2]

Postulates

1.	*a.*	$0 \cdot 0 = 0$	A closed circuit in parallel with a closed circuit is a closed circuit.
	b.	$1 + 1 = 1$	An open circuit in series with an open circuit is an open circuit.
2.	*a.*	$1 + 0 = 0 + 1 = 1$	An open circuit in series with a closed circuit in either order (i.e., whether the open circuit is to the right or left of the closed circuit) is an open circuit.
	b.	$0 \cdot 1 = 1 \cdot 0 = 0$	A closed circuit in parallel with an open circuit in either order is a closed circuit.
3.	*a.*	$0 + 0 = 0$	A closed circuit in series with a closed circuit is a closed circuit.
	b.	$1 \cdot 1 = 1$	An open circuit in parallel with an open circuit is an open circuit.
4.	At any given time either $X = 0$ or $X = 1$.		

A Symbolic Analysis of Relay and Switching Circuits*

*Claude E. Shannon***

I. Introduction

In the control and protective circuits of complex electrical systems it is frequently necessary to make intricate interconnections of relay contacts and switches. Examples of these circuits occur in automatic telephone exchanges, industrial motor-control equipment, and in almost any circuits designed to perform complex operations automatically. In this paper a mathematical analysis of certain of the properties of such networks will be made. Particular attention will be given to the problem of network synthesis. Given certain characteristics, it is required to find a circuit incorporating these characteristics. The solution of this type of problem is not unique and methods of finding those particular circuits requiring the least number of relay contacts and switch blades will be studied. Methods will also be described for finding any number of circuits equivalent to a given circuit in all operating characteristics. It will be shown that several of the well-known theorems on impedance networks have roughly analogous theorems in relay circuits. Notable among these are the delta-wye and star-mesh transformations, and the duality theorem.

The method of attack on these problems may be described briefly as follows: any circuit is represented by a set of equations, the terms of the equations corresponding to the various relays and switches in the circuit. A calculus is developed for manipulating these equations by simple mathematical processes, most of which are similar to ordinary algebraic algorisms. This calculus is shown to be exactly analogous to the calculus of propositions used in the symbolic study of logic. For the synthesis problem the desired characteristics are first written as a system of equations, and the equations are then manipulated into the form representing the simplest circuit. The circuit may then be immediately drawn from the equations. By this method it is always possible to find the simplest circuit containing only series and parallel connections, and in some cases the simplest circuit containing any type of connection.

Our notation is taken chiefly from symbolic logic. Of the many systems in common use we have chosen the one which seems simplest and most suggestive for our interpretation. Some of our phraseology, such as node, mesh, delta, wye, etc., is borrowed from ordinary network theory for simple concepts in switching circuits.

* *Transactions American Institute of Electrical Engineers*, vol. 57, 1938. (Paper number **38-80,** recommended by the AIEE committees on communication and basic sciences and presented at the AIEE summer convention, Washington, D.C., June 20-24, 1938. Manuscript submitted March 1, 1938; made available for preprinting May 27, 1938.)

** Claude E. Shannon is a research assistant in the department of electrical engineering at Massachusetts Institute of Technology, Cambridge. This paper is an abstract of a thesis presented at MIT for the degree of master of science. The author is indebted to Doctor F. L. Hitchcock, Doctor Vannevar Bush, and Doctor S. H. Caldwell, all of MIT, for helpful encouragement and criticism.

Shannon, C.E., "A symbolic analysis of relay and switching circuits", *Transactions of the American Institute of Electrical Engineers*, Vol. 57, 1938, 713-723.

Theorems of Boolean algebra used by Shannon in simplification of relay circuits

$$X + Y = Y + X\,, \tag{1a}$$

$$XY = YX\,, \tag{1b}$$

$$X + (Y + Z) = (X + Y) + Z\,, \tag{2a}$$

$$X(YZ) = (XY)Z\,, \tag{2b}$$

$$X(Y + Z) = XY + XZ\,, \tag{3a}$$

$$X + YZ = (X + Y)(X + Z)\,, \tag{3b}$$

$$1 \cdot X = X\,, \tag{4a}$$

$$0 + X = X\,, \tag{4b}$$

$$1 + X = 1\,, \tag{5a}$$

$$0 \cdot X = 0\,. \tag{5b}$$

[Ref. Shannon 2]

Table I. Analogue Between the Calculus of Propositions and the Symbolic Relay Analysis

Symbol	Interpretation in Relay Circuits	Interpretation in the Calculus of Propositions
X	The circuit X	The proposition X
0	The circuit is closed	The proposition is false
1	The circuit is open	The proposition is true
$X + Y$	The series connection of circuits X and Y	The proposition which is true if either X or Y is true
$X\,Y$	The parallel connection of circuits X and Y	The proposition which is true if both X and Y are true
X'	The circuit which is open when X is closed and closed when X is open	The contradictory of proposition X
=	The circuits open and close simultaneously	Each proposition implies the other

The *Shannon expansion rule* widely used in Switching Theory and Logic Design. In modern notation usually writen as

$$f(x_1,\ldots,x_i,\ldots,x_n) = \bar{x}_i f_0 \oplus x_i f_1,$$

$$f_0 = f(x_1,\ldots,x_i = 0,\ldots,x_n), \quad f_1 = f(x_1,\ldots,x_i = 1,\ldots,x_n)$$

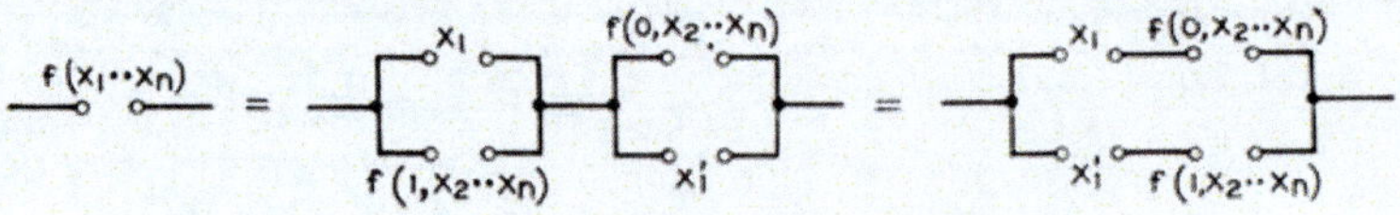

Figure 4. Expansion about one variable

6.2 Review of the Work by Shannon

Boole used symbolic logic to "Rephrase a complicated statement in a simpler, more convenient, form without changing its meaning". Shannon used it to simplify relay circuits.

Main principle

Shannon used the following correspondence between the circuits and logic symbols.

1. If a terminal is open, it has an infinite impedance and the logic value 1 is assigned.
2. For a closed terminal, the impedance is zero, and the logic value 0 is assigned.
3. Negation for a terminal X is defined as the value opposite to the value assigned to X.

The variable X_{ab} connecting two terminals a and b is viewed as a function of time and called hindrance of the two terminal circuit $a-b$. For simplicity, the symbol X_{ab} is often replaced by X.

For the simple relay locking circuit, Shannon uses the relation $X = A(B+X)$ and interprets it as the relay X operates iff either circuit A is closed or both circuit B is closed and relay X is operating.

The analysis and synthesis method for circuits has been expressed in the following way [Shannon, 1938]

Any circuit is represented by a set of equations, the terms of the equations corresponding to the various relays and switches in the circuit.

A calculus is developed for manipulating these equations by simple mathematical processes, most of which are similar to ordinary algebraic algorithms.

For the synthesis problem the desired characteristics are first written as a system of equations, and the equations are then manipulated into the form representing the simplest circuit.

Frequently a function may be written in several ways, each requiring the same minimum number of elements. In such a case the choice of circuit may be made arbitrarily from among these, or from other considerations.

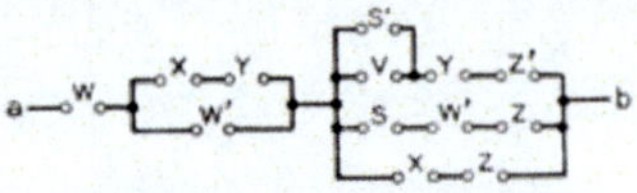

Figure 5. Circuit to be simplified

As an example of the simplification of expressions consider the circuit shown in Figure 5. The hindrance function X_{ab} for this circuit will be:

$$\begin{aligned} X_{ab} &= W + W'(X + Y) + (X + Z)(S + W' + Z)(Z' + Y + S'V) \\ &= W + X + Y + (X + Z)(S + 1 + Z)(Z' + Y + S'V) \\ &= W + X + Y + Z(Z' + S'V) \,. \end{aligned}$$

These reductions were made with 17b using first W, then X and Y as the "X" of 17b. Now multiplying out:

$$\begin{aligned} X_{ab} &= W + X + Y + ZZ' + ZS'V \\ &= W + X + Y + ZS'V \,. \end{aligned}$$

The circuit corresponding to this expression is shown in Figure 6. Note the large reduction in the number of elements.

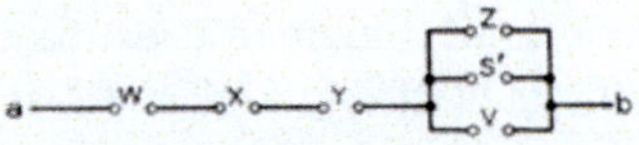

Figure 6. Simplification of figure 5

An example of simplificaiton of a realy circuits by using postulates and theorems of the Boolean algebra.

References

1. A complete bibliography of the literature of symbolic logic is given in the *Journal of Symbolic Logic*, volume 1, number 4, December 1936. Those elementary parts of the theory that are useful in connection with relay circuits are well treated in the two following references.
2. The Algebra of Logic, **Louis Cauturat.** The Open Court Publishing Company.
3. Universal Algebra, **A. N. Whitehead.** Cambridge, at the University Press, volume I, book III, chapters I and II, pages 35-42.
4. **E. V. Huntington,** *Transactions* of the American Mathematical Society, volume 35, 1933, pages 274-304. The postulates referred to are the fourth set, given on page 280.

References in the paper by Shannon in 1938.

The first system for automatic synthesis of switching circuits

6.3 Review by Baylis

An interpretation in terms of relay and switching circuits of the two-valued Boolean algebra (the author calls it an analogue of the classical propositional calculus). The variables are circuits, $= 0$ means that the circuit is closed, $= 1$ means the circuit is open, $+$ signifies a series connection of circuits, juxtaposition or dot a parallel connection.

Switches are so arranged that whenever a circuit x is open, there is a circuit x' which is closed, and whenever x is closed x' is open. Two circuits are equivalent when they are open and close together. The author shows that every function of the algebra, thus interpreted, represents a series-parallel circuit, and each such is represented by a function of the algebra. He shows further that certain types of non-series-parallel circuits are represented by and may be discovered from symmetric functions, and that equations of the algebra may be written to describe the operating characteristics of such circuits.

Baylis, C.A., "Review of A Symbolic Analysis of Relay and Switching Circuits by Claude E. Shannon", *The Journal of Symbolic Logic*, Vol. 4, No. 2, 1939, 103.

C.E. Shannon, E.F. Moore,"Machine aid for switching circuit design", *Proc. IRE*, October 1953, 1348–1351.

Introduction

SOME OPERATIONS which assist in the design of relay circuits or other types of switching circuits can be described in very simple form, and machines can be constructed which perform them more quickly and more accurately than a human being can. It seems possible that machines of this type will be useful to those whose work involves the design of such circuits.

The first system for automatic sysnthesis of switching circuits.

The circuit shown in Fig. 2 is an example of a circuit which was not known to be inefficiently designed until put on the analyzer. It determined in less than two minutes (including the time required to plug the circuit into the plugboard) that one of the contacts shown can be shorted out. How likely would a human being be to solve this same problem in the same length of time?

After the short test has been performed, putting the main control switch in the "open test" position permits the analyzer to perform another analogous test, this time opening the contacts one at a time.

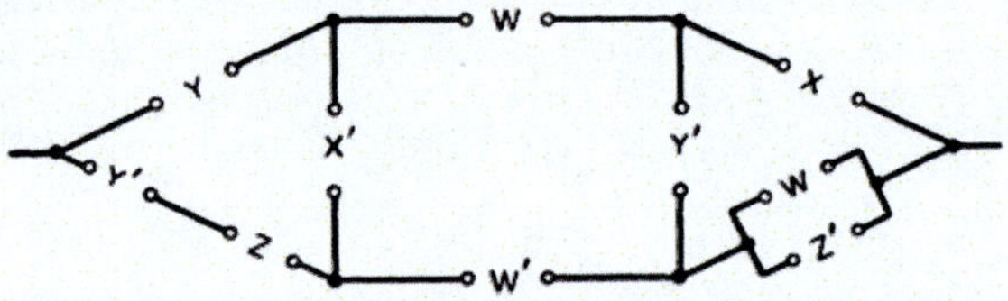

Fig. 2—The relay circuit analyzer was able to simplify this circuit, removing one contact, in less than two minutes total time. Can you do as well?

Conclusions

An interesting feature of the relay circuit analyzer is its ability to deal directly with logical circuits in terms of 3-valued logic. There would be considerable interest in techniques permitting easy manipulation on paper with such a logic, because of its direct application to the design of economical switching circuits. Even though such techniques have not yet been developed, machines such as this can be of value in connection with 3-valued problems.

Ternary Logic

Chapter 7
Switching Theory in USSR

R.S. Stanković & J. Astola: From Boolean Log. to Switch. Circuits & Automata, SCI 335, pp. 141–152.
springerlink.com

Viktor Ivanovič Šestakov

7.1 Pioneering Work by Šhestakov

The ideas suggested by Ehrenfest about an algebra of switching circuits have been explored and elaborated by V.I. Šestakov, a student of V.I. Glivenko, and the results were reported in written form in January 1935, however, as Anovskaja [Anovskaja] states this paper has not been published at the time, and has been used as foundations for the PhD candidate Thesis by Šestakov.

See also [Bažanov], [Bažanov, 2007], and [Kautz].

Šestakov, V.I., *Some Mathematical Methods for Construction and Simplification of Two-element Electrical Networks of Class A*, PhD Dissertation, Lomonosov State University, Moscow, Russia, 1938.

Anovskaja, S.A., (Yanovskaya Sofija Aleksandrovna) "Osnovania matematiki i matematiceskaa logika (Foundations of mathematics and mathematical logic)", Mathematika v SSSR za tridcat let 1917-1947 (Mathematics in the USSR for the thirty years 1917-1947), OGIZ, Moscow and Leningrad, 1948, 9-50.

Bažanov, V.A, "V.I. Shestakov and C. Shannon - Different fates of one brilliant idea authors", *Voprosy Istorii Estestvoznaniya i tehniki*, No. 2, 2005, 112-121.

Bažanov, V.A., *History of Logic in Russia and the USSR - Conceptual Context of University Philosophy*, Canon+, Moscow, 2007.

Kautz, W.H., "A survey and assessment of progress in switching theory and logical design in the Soviet Union", *IEEE Transactions on Electronic Computers*, Vol. EC-15, No. 2, April 1966, 164-204.

The major part of the thesis of Šestakov has been published in the following publications

Šestakov, V. I., "The algebra of two-terminal networks constructed exclusively of two-terminal elements (The algebra of A-networks)", *Avtomatika i Telemekhanika*, Vol. 2, No. 6, 1941, 15-24.

Šestakov, V. I., "The algebra of two-terminal networks constructed exclusively of two-terminal elements (The algebra of A-networks)", *Žurnal Tekh. Fis.*, Vol. 11, No. 6, 1941, 532-549.

with further considerations in

Šestakov, V. I., "A symbolic calculus applied to the theory of electrical relay networks", *Učenye Zapiski Moskowskog Gosudarstvenog Universiteta*, Vol. 73, No. 5, 1944, 45-48.

Šestakov, V. I., "The representation of the characteristic functions of propositions by means of expressions which are realized by relay-contact networks", *Izv. Akad. Nauk.*, Ser. Matem., Vol. 10, 1946, 529-554.

Photos of Šestakov around 1940 at the page 105, and around 1970 at this page are taken from the book by V.A. Bažanov, where they were published for the first time.

Šestakov defended his thesis in the physico-mathematical sciences on September 28, 1938, at the State University Lomonosov, Moscow, Soviet Union.

The University has been decorated by the Order of Lenin.

This means that the Univeristy with such a high distinct, has been free to suport research in areas where other insitutions could hardly do the same.

It should be recalled that in a period of soviet regime, the logic including mathematical logic, has been considered as a bourgeois science coming from the West and not being a subject suitable for a deep scientific study in USSR.

Šestakov studied algebraic properties of *A*-networks, i.e., networks consisting of two-input logic elements. In this order, he introduced the notion of *A*-expressions describing *A*-networks and studied a transition from the algebra of *A*-expressions to the Boolean algebra and equality of these algebras.

For the mathematical background of his study, Šestakov refers to

V.I. Glivenko, *Probability Theory*, 2nd edition, Moscow, 1939, where a definition of the Boolean algebra is given at page 209.

I.I. Žegalkin, M.I. Sludskaja, *Introduction to Analysis*, Učpedgiz, Moscow, 1936.

I.S. Goldštein, *Direct and Inverse Theorems*, ONTI, Moscow, Leningrad, SSSR, 1936.

1941 АВТОМАТИКА И ТЕЛЕМЕХАНИКА № 2

АЛГЕБРА ДВУХПОЛЮСНЫХ СХЕМ, ПОСТРОЕННЫХ ИСКЛЮЧИТЕЛЬНО ИЗ ДВУХПОЛЮСНИКОВ (АЛГЕБРА А-СХЕМ)[1]

Канд. физ.-мат. наук **В. И. Шестаков**

ВЗАИМНО-ОДНОЗНАЧНОЕ СООТВЕТСТВИЕ МЕЖДУ А-СХЕМАМИ И А-ВЫРАЖЕНИЯМИ

А-схемы

В настоящей работе установлено взаимно-однозначное соответствие между А-схемами, т. е. двухполюсными схемами, построенными исключительно из двухполюсников, и А-выражениями, т. е. алгебраическими выражениями, члены которых являются символами двухполюсников и связаны друг с другом лишь двумя операциями: сложения и гармонического сложения.

Установленное взаимно-однозначное соответствие позволяет написать А-выражение по заданному начертанию А-схемы и, наоборот, позволяет начертить А-схему по заданному А-выражению этой схемы и производить всякого рода преобразования А-схем (в частности, и упрощение А-схем) посредством алгебраических преобразований, соответствующих схемам А-выражений.

Далее показано, что алгебра вырожденных А-схем, т. е. таких А-схем, проводимость которых равна 0 или ∞, является алгеброй Буля. Отсюда следует, что для конструирования релейных схем, срабатывающих от заданных сочетаний одновременно или последовательно передаваемых элементарных сигналов, может быть привлечен весь аппарат алгебры Буля.

Двухполюсники X_1 и X_2 называют „соединенными друг с другом" или, что то же, „из двухполюсников X_1 и X_2 построена схема" или „двухполюсники X_1 и X_2 образуют схему", если по крайней мере один полюс одного из них находится в электрическом контакте с полюсом другого из них.

Полюсы двухполюсников, между которыми существует электрический контакт, называют узлами схемы. Узел может быть, в свою очередь, полюсом, т. е. может допускать присоединение к нему других схем или источников электродвижущей силы; но может быть и так, что к узлу дальнейшее присоединение уже не допускается.

Графически узлы мы будем изображать посредством совмещения полюсов, образующих узел, причем узел, являющийся полюсом, будем изображать пустым кружком. Узел же, не являющийся полюсом, будем обозначать черным кружком.

Для иллюстрации приведены фиг. 1, *a* и *b*. На фиг. 1, *b* изображена та же схема, что и на фиг. 1, *a*, но рассматривается она уже не как двухполюсник, а как трехполюсник. В схеме фиг. 1 допускается присоединение проводов не только к полюсам I и II, но также и к узлу III, являющемуся местом последовательного соединения двухполюсников X_1 и X_2. Не допуская совмещения обоих полюсов одного и того же двухполюсника,

[1] Настоящая работа представляет собой краткое изложение части кандидатской диссертации, защищенной автором 28 сентября 1938 г. в Московском ордена Ленина государственном университете им. М. В. Ломоносова.

As it has been pointed out in [Kautz], credit to Šestakov is given by many Soviet authors, as for example, in [PetrovRutkhovskii], although the work of Shannon is also well recognized and referred. See, also [Bažanov], [Bažanov Russian], [Bažanov, 2007].

[Bažanov Russian]
Bažanov, V.A., Očerki sotsialnoj istorii logiki v Rossii
(Essays in the Social History of Logic in Russia),
Izdatelstvo Srednevolžskogo naučnogo centra,
(Ulyanovsk-Simbirsk, Mid-Volga Research Center Publ.), 2002.

[PetrovRutkhovskii]
Petrov, V. V., Rutkhovskii, V. Yu., "Some problems relative to the logical design of circuits and the selection of specifications for high-speed servomechanisms", *Avt. Uprav. i Vych. Tekh.*, Vol. 2, 1959.

The problem has been discussed further by A.M. Gavrilov in a series of papers published between 1943 to 1947. The book by Gavrilov established a basis for further study of switching theory in Soviet Union, and since it has been translated in German, the book had a considerable influence abroad.

Gavrilov, M. A., "The synthesis and analysis of relay-contact networks", *Avtomatika i Telemekhanika*, Vol. 4, 1943.

Gavrilov, M. A., *The Theory of Relay-Contact Networks*, Izdat. Akad. Nauk SSSR, Moscow, 1950 translation in German *Relaisschalttechnik fur Starkund Schwachstromanlagen*, V.E.B. Verlag Technik, Berlin, 1963.

Review of the book in

Zdzislaw Pawlak, *Review of Theory of Relay-Contact Networks. Analysis and Synthesis of the Structure of Relay - Contact Networks*, by M. A. Gavrilov, *The Journal of Symbolic Logic*, Vol. 21, No. 3, September 1956, 331.

7.2 Contributions by Other Researchers

M. A. GAVRILOV. ***Téoriá réléjno-kontaktnyh shém. Analiz i sintéz struktury réléjno-kontaktnyh shém.*** (Theory of relay-contact networks. Analysis and synthesis of the structure of relay-contact networks.). Akadémiá Nauk SSSR, Institut Avtomatiki i Téléméhaniki. Izdatél'stvo Akadémii Nauk SSSR, Moscow-Leningrad 1950, 302 pp.

The book of Gavrilov is to the reviewer's knowledge the first book in the world literature devoted entirely to the synthesis and analysis of relay circuits. The author considers in it a very wide class of problems connected with the subject, for example two- and n-terminal circuits, bridge-type networks, circuits with feedback, sequential switching circuits, networks containing relays with more than one winding, networks with amplitude sensitive-relays, selectors, rectifiers, relays with delayed action, etc. The main part of the book is concerned with synthesis and analysis of series-parallel networks (circuits of class П in the Russian terminology) and networks containing bridge-type connections (circuits of class H in the Russian terminology). Most of these problems are treated with the aid of a more or less modified Boolean algebra. For example, if in sequential switching circuits the switching-on of contacts x_1 and x_2 follows in the order of increasing indices, and the switching-off in the reverse order, then the following laws hold: $x_1 + \bar{x}_2 = 1$, $\bar{x}_1 \cdot x_2 = 0$, $\bar{x}_1 + \bar{x}_2 = \bar{x}_2$, $x_1 \cdot x_2 = x_2$, where the symbols 1, 0, +, $\cdot$, and $\bar{}$ are interpreted in familiar manner.

The book is full of interesting suggestions. Unfortunately they are not developed, and so it is not possible to describe them in a detailed way in a short review. Especially interesting problems arise in connection with the author's method of classification of relay-circuits. However, the author does not go deeper into this matter.

The introduction to the book is devoted to a sketch of the history of the application of Boolean algebra to analysis and synthesis of relay networks in the USSR and other countries.

Forty-one citations of the literature are given, including twenty-eight papers of Russian authors.

The book as a whole represents a valuable manual for relay-circuit designers. The mathematical method used in it seems to the reviewer to be not the best one. The presentation would have gained in clarity, exactitude, and compactness, had the author decided to use the terminology of the theory of graphs instead of the terminology of Boolean algebra. It is also to be remarked that problems are formulated mainly with the aid of examples, which in some cases makes difficulty in an accurate understanding of the author's ideas.

The definition no. 10 on page 279 is circular. ZDZISŁAW PAWLAK

M.A. Gavrilov, "Methods of synthesis of relay-contact networks, *Električéstvo*, No. 2, 1946, 54-59.

M.A. Gavrilov, "Fundamental formulas of the synthesis of relay circuits", *Avtomatika i Telemekhanika*, Vol. 15, 1954, 521-537.

M.A. Gavrilov, "Relay networks with rectifiers", *Avtomatika i Telemakhanika*, Vol. 16, 1955, 328.343.

Reviewed by Z. Pawlak in *The Journal of Symbolic Logic*, Vol. 23, No. 3, September 1958, 367 .

M.A. Gavrilov, "Synthesis of bridge-type relay circuits with given resistance conditions", *Avtomatika i Telemekhanika*, Vol. 14, 1953, 188-198.

M.A. Gavrilov, "Selecting in relay networks paths energizing a given effector", *DoklaDy Akademii Nauk SSSR*, Vol. 87, 1952, 413-416.

M.A. Gavrilov, "Determination of the sequence of operation in relay networks", *Avtomatkai Telemekhanika*, Vol. 13, 1952, 413-416.

Review by Z. Pawlak in
The Journal of Symbolic Logic, Vol. 23, No. 2, June 1958, 234.

This area has been studied by many other researchers, for example,

Povarov, G. N., "Matrix methods of analyzing relay-contact networks in terms of the conditions of non-operation", *Avtomatika i Telemekhanika*, Vol. 15, No. 4, 1954, 332-335.

G.N. Povarov, "On the synthesis of n-terminal relay-contact networks", *Doklady Akademii Nauk SSSR*, Vol. 94, 1954, 1075-1078.

G.N. Povarov, "On the functional separability of Boolean functions",
English translation in
Morris D. Freiedman, Inc., Needham Heights, Mass., USA, 1957, 4 pages.

S.V. Yablonskij, "On the functional completeness in three-valued calculus",
English translation in
Morris D. Freidman, Inc., Needham Heights, Mass., USA, 1957, 3 pages.

S.V. Yablonskij, "Realization of linear functions in *P* (series-parallel)-circuits",
English translation in
Morris D. Friedman, Inc., Needham Heights, Mass., USA, 1957, 4 pages.

Review by Z. Pawlak in
The Journal of Symbolic Logic, Vol. 23, No. 3, September 1958, 367.

Доклады Академии Наук СССР
1954. Том XCIV, № 5

МАТЕМАТИКА

Г. Н. ПОВАРОВ

О ФУНКЦИОНАЛЬНОЙ РАЗДЕЛИМОСТИ БУЛЕВЫХ ФУНКЦИЙ

(Представлено академиком М. В. Келдышем 7 XII 1953)

1. Развитие теории релейных схем (1) ставит задачу исследования систематики булевых функций.

Условия работы однотактных релейных схем выражаются булевыми функциями, аргументы которых изображают состояние приемных элементов (1). Множество всех булевых функций n аргументов описывает все возможные условия работы схемы с n приемными элементами.

Однако условия работы, обычно встречающиеся на практике, как правило, принадлежат к некоторым особым множествам (классам) булевых функций. Благодаря этому для их схемной реализации требуется значительно меньше контактов, чем для реализации произвольных булевых функций, которые, вообще говоря, имеют весьма сложный вид.

Представляется важным выделить классы простых булевых функций и исследовать соотношения между такими классами.

К числу классов простых булевых функций принадлежит класс функционально разделимых функций, описанных К. Э. Шенноном (3). В настоящей работе предлагается критерий функциональной разделимости булевых функций и с его помощью разыскивается соотношение между функционально разделимыми и симметрическими булевыми функциями.

2. Функция n переменных $f(x_1, x_2, \ldots, x_n)$ называется функционально разделимой, если существуют такие функции $g(z_1, z_2, \ldots, z_{n-s+1})$ и $h(z_2, z_2, \ldots, z_s)$, что

$$1 < s < n \text{ и } f(x_1, x_2, \ldots, x_n) = g(h(y_1, y_2, \ldots, y_s), y_{s+1}, \ldots, y_n),$$

где $y_1, y_2, \ldots, y_s$ — некоторое подмножество переменных $x_1, x_2, \ldots, x_n$, а $y_{s+1}, y_{s+2}, \ldots, y_n$ — остальные из этих переменных*.

Иными словами, функционально разделимая функция представима в виде подстановки функции не менее двух переменных в функцию не менее двух же переменных.

По определению функционально разделимые функции возможны лишь начиная с $n = 3$. Тривиальным примером функционально разделимых функций являются суммы и произведения, составленные из переменных и их отрицаний.

Реализация функции $f = g(h, \ldots)$ сводится к реализации функций g и h и подстановке схемы для h в схему для g («блочный метод

* Это определение несколько отличается от определения К. Э. Шеннона. Он требовал, чтобы $1 < s < n - 1$.

Information on publications in this subject by Gavrilov, Povarov, Šestakov and their associates can be found in the following publications

Kautz, W.H., "A survey and assessment of progress in switching theory and logical design in the Soviet Union", *IEEE Trans. on Electronic Computers*, Vol. EC-15, No. 2, April 1966, 164-204.

Kung, G., "Bibliography of soviet work in the field of mathematical logic and foundations of mathematics from 1917-1957", *Notre Dame Journal of Formal Logic*, Vol. 3, No. 1, January 1962, 1-40.

Stanković, R.S., Sasao, T., Astola, J.T., *Publications in the First Twenty Years of Switching Theory and Logic Design*, TICSP Series #14, ISBN 952-15-0679-2, Tampere University of Technology, Tampere, Finland, 2001.

As noticed in [Kautz], related work by V.A. Rozenberg, and M.G. Tsimbalistii have not been widely reported.

Rozenberg, V. A., "The problem of blocking and transformation of contact groups", *Avtomatika i Telemekhanika*, No. 1, 1940, 47-54.

Rozenberg, V. A., "Some properties of relay selection", *Avtomatika i Telemekhanika*, No. 1, 1939, 37-48.

Tsimbalistii, M. G., "The economical design of relay networks", *Trudy Leningradskoi Eksperimentalnoi Elektrotekhniceskoi Labor.*, Vol. 8, 1928, 19-31.

Tsimbalistii, M. G., "The problem of simplification in the design of relay networks", *Trudy Leningradskogo Eksper. Elektrotekhniceskoi Labor.*, No. 8, 1928, 19-31.

Work by Šestakov, Gavrilov and their followers have been reviewed by Ljapunov in

Ljapunov, A.A., "Matematiceskie isledovania sviazannye s eksploatacei elektronnukh vycislitelnih main", *Matematika v SSSR za sorok let*, Moscow, 1959.

Chapter 8
Switching Theory in Japan

R.S. Stanković & J. Astola: From Boolean Log. to Switch. Circuits & Automata, SCI 335, pp. 153–166.
springerlink.com

Akira Nakashima

8.1 Akira Nakashima

1935 1. Nakashima, A., "The theory of relay circuit engineering", *Nichiden Geppo* (Nippon Electric) Nippon Electrical Company (NEC)}, November 1934 - September 1935, (in Japanese)

1936 2. Nakashima, A., "Synthesis theory of relay networks", *Journal of the Institute of Telegraph and Telephone Engineers of Japan*, No. 150, September 1935, 731-752. English version in *Nippon Electrical Communication Engineering*, No. 3, May 1936, 197-226.

1937 3. Nakashima, A., "Reziprozitaetsgesetze", *Nichiden Geppo - Nippon Electric*, Nippon Electrical Company (NEC), January 1936 (in Japanese).

1938 4. Nakashima, A., "Some properties of the group of simple partial paths in the relay circuit", *Journal of the Institute of Telephone and Telegraph Engineers of Japan*, January, February, and March 1936, 88 - 95, English translation in *Nippon Electrical Communication Engineering*, March 1937, 70 -71.

5. Nakashima, A., Hanzawa, M., "The theory of equivalent transformation of simple partial paths in the relay circuit (Part 1), *Journal of the Institute of Electrical Communication Engineers of Japan*, No. 165, October 28, 1936, published December 1936. Condensed English version of parts 1 and 2 in *Nippon Electrical Communication Engineering*, No. 9, February 1938, 32-39.

6. Nakashima, A., Hanzawa, M., "The theory of equivalent transformation of simple partial paths in the relay circuit (Part 2), *Journal of the Institute of Electrical Communication Engineers of Japan*, No. 167, December 14, 1936, published in February 1937. Condensed English version of parts 1 and 2 in *Nippon Electrical Communication Engineering*, No. 9, February 1938, 32-39.

1938 7. Nakashima, A., "The theory of four-terminal passive networks in relay circuit", *Journal of the Institute of Electrical Communication Engineers of Japan*, April 1937, English summary in *Nippon Electrical Communication Engineering*, No. 10, April 1938, 178-179.

8. Nakashima, A., "Algebraic expressions relative to simple partial paths in the relay circuits", *Journal of the Institute of Electrical Communication Engineers of Japan*, No. 173, August 1937, in Japanese. Condensed English translation of about half the length of the original paper in *Nippon Electrical Communication Engineering*, No. 12, September 1938, 310-314. Section V, "Solutions of acting impedance equations of simple partial paths".

9. Nakashima, A., "The theory of two-point impedance of passive networks in the relay circuit (Part 1)", *Journal of the Institute of Electrical Communication Engineers of Japan*, No. 177, December 1937. Reduced version of part 1 and part 2 in *Nippon Electrical Communication Engineering*, No. 13, November 1938, 405-412.

10. Nakashima, A., "The theory of two-point impedance of passive networks in the relay circuit (Part 2)", *Journal of the Institute of Electrical Communication Engineers of Japan*, No. 178, January 1938. Reduced version of part 1 (above) and 2 in Nippon Electrical Communication Engineering, No. 13, November 1938, 405-412.

11. Nakashima, A., "The transfer impedance of four-terminal passive networks in the relay circuit", *Journal of the Institute of Electrical Communication Engineers of Japan*, No. 179, February 1938. Condensed English version in *Nippon Electrical Communication Engineering*, No. 14, December 1938, 459-466.

1941 12. Nakashima, A., Hanzawa, M., "Expansion theorem and design of two-terminal relay networks (Part 1)", *Journal of the Institute of Electrical Communication Engineers of Japan*, No. 206, May 1940. Condensed English version in *Nippon Electrical Communication Engineering* , No. 24, April 1941, 203-210.

13. Nakashima, A., Hanzawa, M., "Expansion theorem and design of two-terminal relay networks (Part 2)", *Journal of the Institute of Electrical Communication Engineers of Japan*, No. 209, August 1940. Condensed English version in *Nippon Electrical Communication Engineering*, No. 26, October 1941, 53-57.

14. Nakashima, A., "Theory of relay circuit", *Journal of the Institute of Electrical Communication Engineers of Japan*, No. 220, March 1941, 9 -12. No English translation of this paper.

The list of publicaitons by Nakashima and Hanzawa has been compiled by Prof. Akihiko Yamada.

Yamada, A., "History of research on switching theory in Japan", IEEJ, C, Vol. 123, No.1, 2003, 1.

Yamada, A., "History of research on switching theory in Japan", *IEEJ Trans. FM*, Vol. 124, No. 8, 2004, 720-726, (in Japanese).

For more details about the work of Akira Nakashima, see

Stanković, R.S., Astola, J.T., (eds.), Reprints from the Early Days of Information Sciences, On the Contributions of Akira Nakashima to Switching Theory, TICSP Series # 40, 2008, ISSN 1456-2774, ISBN 978-952-15-1980-2, 238 pages.

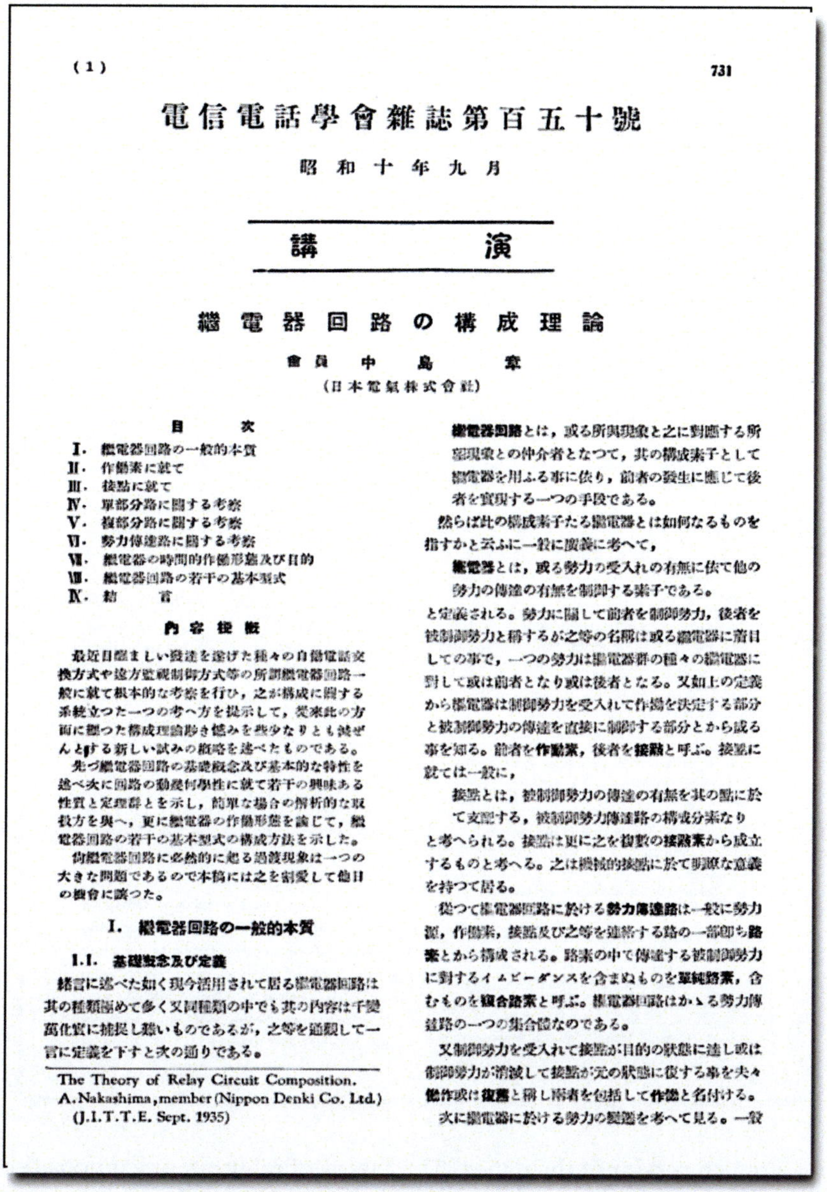

(1) 731

電信電話學會雜誌第百五十號

昭和十年九月

講　演

繼電器回路の構成理論

會員　中島　章
(日本電氣株式會社)

目　次

I. 繼電器回路の一般的本質
II. 作働素に就て
III. 接點に就て
IV. 單部分路に關する考察
V. 複部分路に關する考察
VI. 勢力傳達路に關する考察
VII. 繼電器の時間的作働形態及び目的
VIII. 繼電器回路の若干の基本型式
IX. 結　言

内容梗概

最近目醒ましい發達を遂げた種々の自働電話交換方式や遠方監視制御方式等の所謂繼電器回路一般に就て根本的な考察を行ひ，之が構成に關する系統立つた一つの考へ方を提示して，從來此の方面に缺つた構成理論缺き憾みを些少なりとも減ぜんとする新しい試みの概略を述べたものである。

先づ繼電器回路の基礎概念及び基本的な特性を述べ次に回路の動幾何學性に就て若干の興味ある性質と定理群とを示し，簡單な場合の解析的な取扱方を與へ，更に繼電器の作働形態を論じて，繼電器回路の若干の基本型式の構成方法を示した。

尙繼電器回路に必然的に起る過渡現象は一つの大きな問題であるので本稿には之を割愛して他日の機會に讓つた。

I. 繼電器回路の一般的本質

1.1. 基礎概念及び定義

緒言に述べた如く現今活用されて居る繼電器回路は其の種類極めて多く又同種類の中でも其の内容は千變萬化實に捕捉し難いものであるが，之等を通觀して一言に定義を下すと次の通りである。

繼電器回路とは，或る所與現象と之に對應する所望現象との仲介者となつて，其の構成素子として繼電器を用ふる事に依り，前者の發生に應じて後者を實現する一つの手段である。

然らば此の構成素子たる繼電器とは如何なるものを指すかと云ふに一般に廣義に考へて，

繼電器とは，或る勢力の受入れの有無に依て他の勢力の傳達の有無を制御する素子である。

と定義される。勢力に關して前者を制御勢力，後者を被制御勢力と稱するが之等の名稱は或る繼電器に着目しての事で，一つの勢力は繼電器群の種々の繼電器に對して或は前者となり或は後者となる。又如上の定義から繼電器は制御勢力を受入れて作働を決定する部分と被制御勢力の傳達を直接に制御する部分とから成る事を知る。前者を**作働素**，後者を**接點**と呼ぶ。接點に就ては一般に，

接點とは，被制御勢力の傳達の有無を其の點に於て支配する，被制御勢力傳達路の構成分素なり

と考へられる。接點は更に之を複數の**接點素**から成立するものと考へる。之は機械的接點に於て明瞭な意義を持つて居る。

從つて繼電器回路に於ける**勢力傳達路**は一般に勢力源，作働素，接點及び之等を連絡する路の一部即ち**路素**とから構成される。路素の中で傳達する被制御勢力に對するインピーダンスを含まぬものを**單純路素**，含むものを**複合路素**と呼ぶ。繼電器回路はかゝる勢力傳達路の一つの集合體なのである。

又制御勢力を受入れて接點が目的の狀態に達し或は制御勢力が消滅して接點が元の狀態に復する事を夫々**働作**或は**復舊**と稱し兩者を包括して**作働**と名付ける。

次に繼電器に於ける勢力の變遷を考へて見る。一般

The Theory of Relay Circuit Composition.
A. Nakashima, member (Nippon Denki Co. Ltd.)
(J.I.T.T.E. Sept. 1935)

A. Nakashima, "The theory of relay circuit composition", *The Journal of the Institute of Telephone and Telegraph Engineers of Japan*, No. 150, September 1935, 731-752.

First paper ever published on Switching Theory.

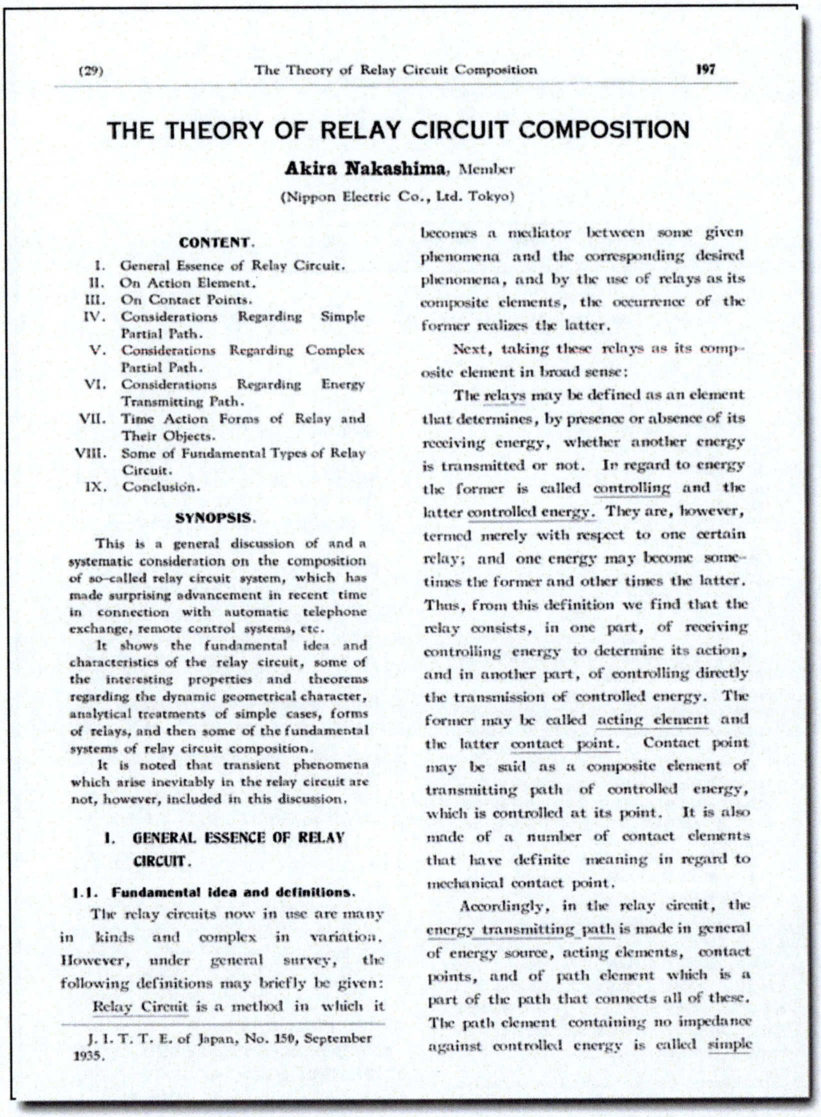

(29) The Theory of Relay Circuit Composition 197

THE THEORY OF RELAY CIRCUIT COMPOSITION

Akira Nakashima, Member

(Nippon Electric Co., Ltd. Tokyo)

CONTENT.

I. General Essence of Relay Circuit.
II. On Action Element.
III. On Contact Points.
IV. Considerations Regarding Simple Partial Path.
V. Considerations Regarding Complex Partial Path.
VI. Considerations Regarding Energy Transmitting Path.
VII. Time Action Forms of Relay and Their Objects.
VIII. Some of Fundamental Types of Relay Circuit.
IX. Conclusion.

SYNOPSIS.

This is a general discussion of and a systematic consideration on the composition of so-called relay circuit system, which has made surprising advancement in recent time in connection with automatic telephone exchange, remote control systems, etc.

It shows the fundamental idea and characteristics of the relay circuit, some of the interesting properties and theorems regarding the dynamic geometrical character, analytical treatments of simple cases, forms of relays, and then some of the fundamental systems of relay circuit composition.

It is noted that transient phenomena which arise inevitably in the relay circuit are not, however, included in this discussion.

I. GENERAL ESSENCE OF RELAY CIRCUIT.

1.1. Fundamental idea and definitions.

The relay circuits now in use are many in kinds and complex in variation. However, under general survey, the following definitions may briefly be given:

Relay Circuit is a method in which it becomes a mediator between some given phenomena and the corresponding desired phenomena, and by the use of relays as its composite elements, the occurrence of the former realizes the latter.

Next, taking these relays as its composite element in broad sense:

The relays may be defined as an element that determines, by presence or absence of its receiving energy, whether another energy is transmitted or not. In regard to energy the former is called controlling and the latter controlled energy. They are, however, termed merely with respect to one certain relay; and one energy may become sometimes the former and other times the latter. Thus, from this definition we find that the relay consists, in one part, of receiving controlling energy to determine its action, and in another part, of controlling directly the transmission of controlled energy. The former may be called acting element and the latter contact point. Contact point may be said as a composite element of transmitting path of controlled energy, which is controlled at its point. It is also made of a number of contact elements that have definite meaning in regard to mechanical contact point.

Accordingly, in the relay circuit, the energy transmitting path is made in general of energy source, acting elements, contact points, and of path element which is a part of the path that connects all of these. The path element containing no impedance against controlled energy is called simple

J. I. T. T. E. of Japan, No. 150, September 1935.

English version of the paper "The theory of relay circuit composition", *Nippon Electrical Communication Engineering*, No. 3, May 1936, 197-226.

Nakashima first did an extensive analysis of many case studies of relay networks trying to formulate a unified design theory for such networks. He considered impedances of relay contacts as two-valued variables, and used logic *OR* and *AND* operations to represent their series and parallel connections, respectively. Due to that, he formulated a related theory of relay networks by introducing and exploiting some algebraic relations that represent a basis of switching theory. For instance, he defined with his associate Masao Hanyawa the rules that are nowadays called *De Morgan duality expressions*. These results Nakashima presented without using a symbolic notation in a series of articles in the monthly journal of NEC entitled *Theory and Practice of Relay Engineering*.

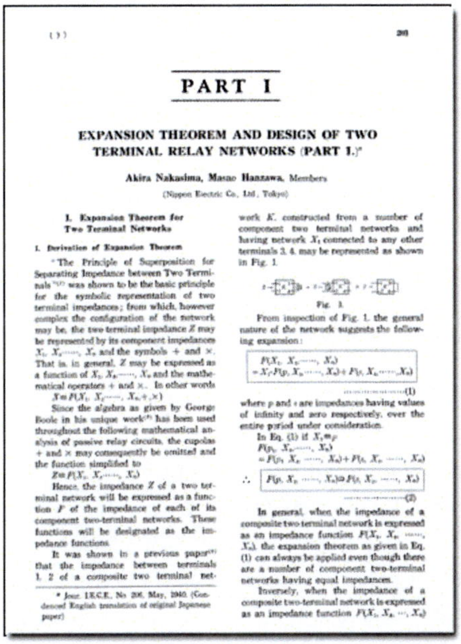

(1) 201

PART I

EXPANSION THEOREM AND DESIGN OF TWO TERMINAL RELAY NETWORKS (PART I.)*

Akira Nakasima, Masao Hanzawa, Members

(Nippon Electric Co., Ltd., Tokyo)

1. Expansion Theorem for Two Terminal Networks

1. Derivation of Expansion Theorem

"The Principle of Superposition for Separating Impedance between Two Terminals"(1) was shown to be the basic principle for the symbolic representation of two terminal impedances; from which, however complex the configuration of the network may be, the two terminal impedance Z may be represented by its component impedances $X_1, X_2, \cdots, X_n$ and the symbols + and ×. That is, in general, Z may be expressed as a function of $X_1, X_2, \cdots, X_n$ and the mathematical operators + and ×. In other words

$Z = F(X_1, X_2, \cdots, X_n, +, \times)$

Since the algebra as given by George Boole in his unique work(2) has been used throughout the following mathematical analysis of passive relay circuits, the cupolas + and × may consequently be omitted and the function simplified to

$Z = F(X_1, X_2, \cdots, X_n)$

Hence, the impedance Z of a two terminal network will be expressed as a function F of the impedance of each of its component two-terminal networks. These functions will be designated as the impedance functions.

It was shown in a previous paper(3) that the impedance between terminals 1, 2 of a composite two terminal network K, constructed from a number of component two terminal networks and having network X_1 connected to any other terminals 3, 4, may be represented as shown in Fig. 1.

Fig. 1

From inspection of Fig. 1, the general nature of the network suggests the following expansion:

$$F(X_1, X_2, \cdots, X_n) = X_1 \cdot F(p, X_2, \cdots, X_n) + F(s, X_2, \cdots, X_n) \quad \cdots(1)$$

where p and s are impedances having values of infinity and zero respectively, over the entire period under consideration.

In Eq. (1) if $X_1 = p$

$F(p, X_2, \cdots, X_n)$
$= F(p, X_2, \cdots, X_n) + F(s, X_2, \cdots, X_n)$

$$\therefore \quad F(p, X_2, \cdots, X_n) \supseteq F(s, X_2, \cdots, X_n) \quad \cdots(2)$$

In general, when the impedance of a composite two terminal network is expressed as an impedance function $F(X_1, X_2, \cdots, X_n)$, the expansion theorem as given in Eq. (1) can always be applied even though there are a number of component two-terminal networks having equal impedances.

Inversely, when the impedance of a composite two-terminal network is expressed as an impedance function $F(X_1, X_2, \cdots, X_n)$

* Jour. I.E.C.E., No. 206, May, 1940. (Condensed English translation of original Japanese paper)

論 文 欄

継電器回路に於けるインピーダンス函数の展開定理と二端子回路網の設計理論 (其の一)

I. 緒 言

Chapter 9
Switching Theory in Europe

R.S. Stanković & J. Astola: From Boolean Log. to Switch. Circuits & Automata, SCI 335, pp. 167–181.
springerlink.com

Mochinori Goto

8.3 Continuation of Research

Work in this area has been continued by many researchers in Japan. The following is an example.

Goto, M., "Applications of logical equations to the theory of relay contact networks", *Electric Soc. of Japan*, Vol. 69, April 1949, 125-130.

(13) 69巻 電氣學會雜誌 · 720號 125

(3) L. S. Nergard & B. Salzberg: Resonant Impedance of Transmission Line. I. R. E. 27, 579 (1939)
(4) [illegible]
(5) [illegible]
(6) E. Karplus: "Component of U. H. F. Field Meter," Electronics Nov., 127 (1946)
(7) D. A. Quarles: "Radar System Consideration." E E. 65, 209 (1946)
(8) C. N. Works: "Resonant Cavities for Dielectric Measurements." J. A. Phys. 18, 605 (1947)
"A Resonant Cavity Method for Measuring Dielectric Properties at Ultra-High Frequencies." I. R. E. 31 245 (1943)

論理數學方程式の繼電器回路網理論への應用 *

正員 後藤以紀 (電氣試驗所)

1. 概説

論理數學とは，論理學上の命題，條件等の論理的關係を式で表わしてその演算を行う數學で，その代數的部門は遠く G. W. Leibniz に端緒を見ると言われて居り，Boole 代數とか命題計算とか呼ばれている論理代數が，G. Boole により作られ(1847 年) E. Schröder, B. A. W. Russel, D. Hilbert 氏等を經て發展され現代に至つている。[illegible]

[illegible]

2. 二値論理數學の基礎的公式

A, *B*, ……*Z* なる文字は論理學としては命題や條件を表わし，それの眞なるか僞なるかに從つて 1 とか 0 とかの値を採ると定め，これを命題の眞理値と言う。[illegible]

$$F_\tau = F(t-\tau) \quad \text{(2·1)}$$

[illegible] *F* の否定を ～*F* で表わし次の如く定義する。

$$\sim F = 1 - F \quad \text{(2·2)}$$

即ち回路の開閉は *F* と ～*F* とは互に反對である。

次に，命題 *A* と *B* との論理積を *A*·*B* 又は *AB* で表わし「*A* と *B* と」の意味である。*A* と *B* とが何方共に眞 (即ち 1) なる場合にのみ *AB* が眞なりと定める。[illegible]

$$AB = \min(A, B) \quad \text{(2·3)}$$

で，接點 *A*, *B* が直列に入つているものの動作が表わされる。

次に，*A* と *B* との論理和を *A*∨*B* で表わし「*A* か *B* か」の意味である。少くとも *A* か *B* か，いずれかが眞なる場合に *A*∨*B* が眞なりと定める。即ち

$$A \vee B = \max(A, B) \quad \text{(2·4)}$$

で，接點 *A*, *B* が並列に入つているものの動作が表わされる。

次に「命題 *A* が眞なる場合には必ず *B* が眞である」ことを表わすのに *A*→*B* と記し，[illegible]と言う。即ち

$$A \to B = (\sim A) \vee B \quad \text{(2·5)}$$

* Application of Logical Mathematics to the Theory of Relay Networks. By Motinori GOTO, Member (Electrotechnical Laboratory). [illegible]

Chapter 9
Switching Theory in Europe

R.S. Stanković & J. Astola: From Boolean Log. to Switch. Circuits & Automata, SCI 335, pp. 167–181.
springerlink.com

Johanna Piesch

9.1 Johanna Piesch

The name of the author varies from publication to publication, e.g., Hanna, Hansi, or Johanna [ChurchArchiv].

1939 Piesch, H., "Begriff der allgemeinen Schaltungstechnik", (Principles of the general switching technique), *Archiv für Elektrotechnik*, Berlin, E.T.Z. Verlag, Vol. 33, Heft 10, 1939, 672-686, in German.

Piesch, H., "Uber die Vereinfachung von allgemeinen Schaltungen", (On the simplification of general switching networks), *Archiv für Elektrotechnik*, Berlin, E.T.Z. Verlag, 1939, Vol. 33, Heft 11, 733746, in German.

1951 Piesch, Johanna, "Systematik der automatischen Schaltungen", (The systematism of automatic switching systems), O.T.F., 5. Jahrgang, Heft 3/4, Springer Verlag Wien, Marz-April 1951, 29-43, in German.

1955 Piesch, J., "Matrices in the switching circuit algebra for designing relay controlled networks", *Arch. Elektrichen. Obertr.*, Vol. 9, No. 10, October 1955, 460-468, in German.

1956 Piesch Johanna, "Analytic representation of active fourpoles", *Archiv der Elektrischen Ubertragung*, Vol. 10, October 1956, 429-437, in German.

1957 Piesch, J., "Contributions to the modern algebra of switching circuits", *Sci. Elec*., Vol. 3, No. 1, 1957, 16-25, in German.

In first two references, she considers circuits (switches) with any finite number of states (positions) and under the assumption that not all of them necessarily have the same number of states. Switches are denoted by symbols, as a,b,c, etc., with different positions indicated by subscripts. In this way, the symbol a_i will denote that the switch a is in the position i.

Capital letters are used to express other propositions as results caused by certain assignments of states, thus, to describe the outputs of the network. The operations of addition, multiplication, and inverse, corresponding to the disjunction, conjunction, and negation, are used to form expressions of the algebra built on the propositions denoted by symbols.

HANSI PIESCH. *Begriff der allgemeinen Schaltungstechnik.* **Archiv für Elektrotechnik,** vol. 33 (1939), pp. 672–686.

HANSI PIESCH. *Über die Vereinfachung von allgemeinen Schaltungen.* Ibid., pp. 733–746.

These early papers about switching algebra are of very considerable historical interest. As his sources the author refers to unpublished work of O. Plechl (Vienna) and to Nakasima and Hanzawa XVIII 346(1). Like Nakasima and Hanzawa, he fails to observe that the algebra is in fact Boolean (or better, propositional) algebra. And various ideas of the nineteenth-century algebra of logic are worked out afresh, and evidently independently.

Not only two-position switches but switches having any finite number of positions are dealt with, and not necessarily all of them the same number of positions. Letters such as $a, b, c, \ldots$ are used for switches, and different positions of a switch are indicated by subscripts. In effect a_0 is used to express the proposition that the switch a is in its first position (the position in which its first contact is closed), and b_3 that the switch b is in its fourth position, and so on. Capital letters may be used to express other propositions, e.g. that a certain light or a certain motor is turned on. And the expressions of the algebra are built up out of these propositional letters by means of the signs for multiplication (i.e., conjunction), addition (i.e., disjunction), and the inverse (i.e., negation). The conditions given in any particular problem will be expressed as conditions on the propositional letters in this notation; and besides the special conditions of a particular problem we must always take into account also all

248 REVIEWS

conditions of the form

$$\bar{a}_i \equiv a_0 + a_1 + \ldots + a_{i-1} + a_{i+1} + \ldots + a_{n-1},$$

where a is any n-position switch.

The author's system is propositional algebra rather than propositional calculus, since only equivalences are asserted. It is, moreover, ordinary two-valued propositional algebra, since every expression of the algebra has (in effect) one or other of two truth-values, truth or falsehood. And it is indeed the reviewer's impression that the author's method of dealing with multiple-position switches can be expected to work more smoothly than proposals which have sometimes been made to use many-valued propositional calculus for this.

Errata: in line 14 from the bottom of page 674, it seems probable that a_i was meant to have a bar over it; at several places on page 675, a_l is evidently a misprint for a_1, and ce a misprint for c_e.

ALONZO CHURCH

Church, A., "Archiv für Electrotechnik- review of papers", *The Journal of Symbolic Logic*, Vol. 30, No. 2, 1952, 247-248.

In her publications 1939 Piesch, did an extension of the work by Nakashima.

Methods by Piesch were used in Gilbert, E.N., "*N*- terminal switching circuits", *The Bell System Technical Journal*, Vol. 30, 1951, 668-688.

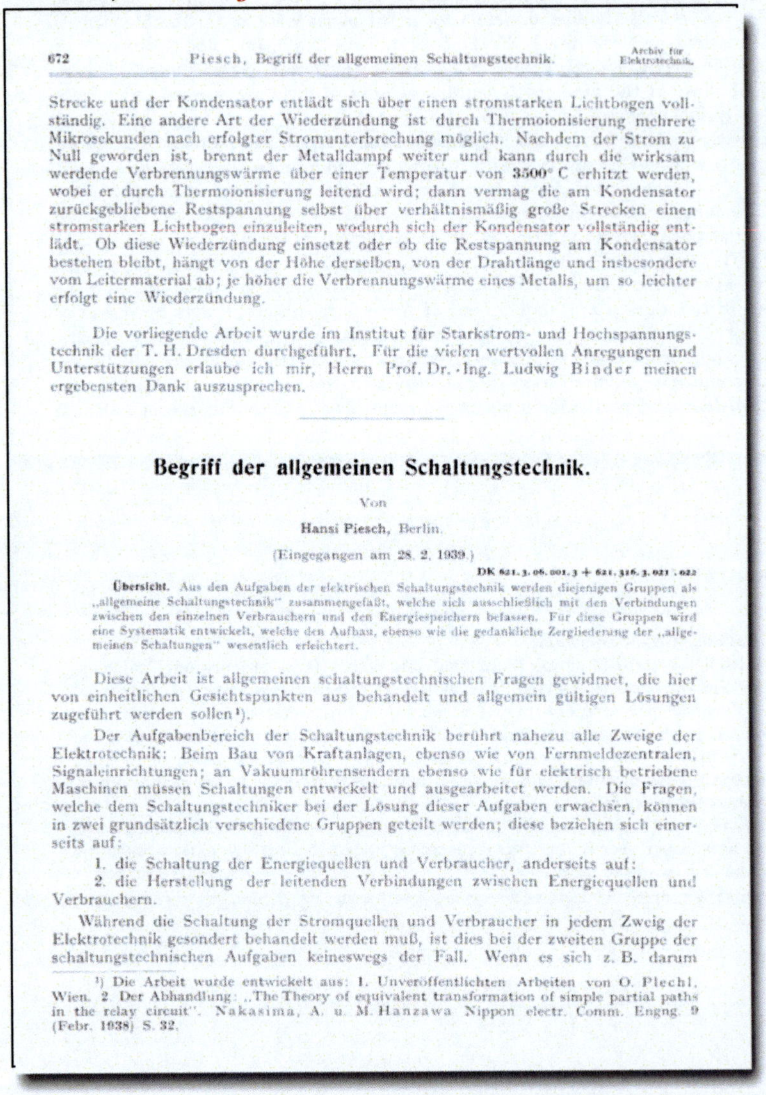

Strecke und der Kondensator entlädt sich über einen stromstarken Lichtbogen vollständig. Eine andere Art der Wiederzündung ist durch Thermoionisierung mehrere Mikrosekunden nach erfolgter Stromunterbrechung möglich. Nachdem der Strom zu Null geworden ist, brennt der Metalldampf weiter und kann durch die wirksam werdende Verbrennungswärme über einer Temperatur von **3500°** C erhitzt werden, wobei er durch Thermoionisierung leitend wird; dann vermag die am Kondensator zurückgebliebene Restspannung selbst über verhältnismäßig große Strecken einen stromstarken Lichtbogen einzuleiten, wodurch sich der Kondensator vollständig entlädt. Ob diese Wiederzündung einsetzt oder ob die Restspannung am Kondensator bestehen bleibt, hängt von der Höhe derselben, von der Drahtlänge und insbesondere vom Leitermaterial ab; je höher die Verbrennungswärme eines Metalls, um so leichter erfolgt eine Wiederzündung.

Die vorliegende Arbeit wurde im Institut für Starkstrom- und Hochspannungstechnik der T. H. Dresden durchgeführt. Für die vielen wertvollen Anregungen und Unterstützungen erlaube ich mir, Herrn Prof. Dr.-Ing. Ludwig Binder meinen ergebensten Dank auszusprechen.

Begriff der allgemeinen Schaltungstechnik.

Von

Hansi Piesch, Berlin.

(Eingegangen am 28. 2. 1939.)

DK 621.3.06.001.3 + 621.316.3.021 .022

Übersicht. Aus den Aufgaben der elektrischen Schaltungstechnik werden diejenigen Gruppen als „allgemeine Schaltungstechnik" zusammengefaßt, welche sich ausschließlich mit den Verbindungen zwischen den einzelnen Verbrauchern und den Energiespeichern befassen. Für diese Gruppen wird eine Systematik entwickelt, welche den Aufbau, ebenso wie die gedankliche Zergliederung der „allgemeinen Schaltungen" wesentlich erleichtert.

Diese Arbeit ist allgemeinen schaltungstechnischen Fragen gewidmet, die hier von einheitlichen Gesichtspunkten aus behandelt und allgemein gültigen Lösungen zugeführt werden sollen[1]).

Der Aufgabenbereich der Schaltungstechnik berührt nahezu alle Zweige der Elektrotechnik: Beim Bau von Kraftanlagen, ebenso wie von Fernmeldezentralen, Signaleinrichtungen; an Vakuumröhrensendern ebenso wie für elektrisch betriebene Maschinen müssen Schaltungen entwickelt und ausgearbeitet werden. Die Fragen, welche dem Schaltungstechniker bei der Lösung dieser Aufgaben erwachsen, können in zwei grundsätzlich verschiedene Gruppen geteilt werden; diese beziehen sich einerseits auf:

1. die Schaltung der Energiequellen und Verbraucher, anderseits auf:
2. die Herstellung der leitenden Verbindungen zwischen Energiequellen und Verbrauchern.

Während die Schaltung der Stromquellen und Verbraucher in jedem Zweig der Elektrotechnik gesondert behandelt werden muß, ist dies bei der zweiten Gruppe der schaltungstechnischen Aufgaben keineswegs der Fall. Wenn es sich z. B. darum

[1]) Die Arbeit wurde entwickelt aus: 1. Unveröffentlichten Arbeiten von O. Plechl, Wien. 2. Der Abhandlung: „The Theory of equivalent transformation of simple partial paths in the relay circuit". Nakasima, A. u. M. Hanzawa Nippon electr. Comm. Engng. 9 (Febr. 1938) S. 32.

Reference to Nakashima and Hanzawa

Otto Plechl

9.2 Vienna School in Switching

In her work Johanna Piesch refers to Nakashima and Hanzawa, and to an unpublished paper by the Austrian researcher Otto Plechl from Vienna.

This author published also latter some interesitng work in this area.

1946 Plechl, O., "Zur Ermittlung elektrischer Kontaktschaltungen", E.u.M., Wien 1946, H. 1/2, 34-38.

Plechl, O., Duschek, A., "Grundzuge einer Algebra der elektrischen Schaltungen", O. Ing.-Archiv, Springer Verlag Wien, 1946, Bd I, H.3, 203-230.

1956 Plechl, O,. with the help by Rieder, W., *Elektromechanische Schaltungenund Schaltgeräte - eine Einführung in Theorie und Berechnung*, Issue Erg. u. bearb., Pulished by von Werner Rieder Erschienen, Wien- Springer, 1956, pages 224.

Otto Plechl worked with his assoicate A. Duschek from Vienna.

For O. Plechl, A. Duschek, R. Edler, see

Piesch, J., Sequenz , H., "Austrian pioneers of the theory of switching circuits", *Elektrotech. u. Maschinenbau,* Vol. 75, No. 9, 1958, 241-245, in German.

O. Plechl und A. Duschek: Grundzüge einer Algebra der elektrischen Schaltungen. 203

von Schönbach bei Eger. Neues Jb. f. Mineralogie **84**, Abt. B, 90—116 (1940). — Einiges über die Untersuchung der geoelektrischen Blitzgefährdung. Gerlands Beitr. Geophys. **57**, 65—108 (1940). — Die Möglichkeit einer geoelektrischen Fehlmutung, mit besonderer Berücksichtigung der bei der Funkmutung gegebenen Voraussetzungen. Elektr. im Bergb. **16**, Nr. 3, 37—43 (1941). — La smorzamento delle onde Hertziane attraverso i conduttori geologici. Geofisica pura e applicata **4**, Nr. 1, 15—37 (1942). — und Heinrich Forejt: Nachweis steil einfallender elektrischer Diskontinuitätsflächen im Untergrunde durch Funkmutung. Hochfrequenztechn. u. E.-A. **59**, Nr. 2, 41—45 (1942). — und Heinrich Forejt: Die Anwendung des Druckindikators in der Funkmutung. Z. Geophysik **17**, Nr. 5/6, 217—225 (1941/42). — Die Aussichten der Reflexionsmethode in der Funkmutung. Glückauf **79**, Nr. 27/28, 336—340 (1943). — Wassersuche durch Funkmutung. Wasserkr. u. Wasserwirtsch. **38**, Nr. 9, 209—215 (1943). — Mitteilung über die im Blitzversuchsfeld Absroth im Jahre 1941 durchgeführten Arbeiten. Gerlands Beitr. Geophys. **59**, Nr. 3/4, 306—330 (1943). — Nachweis von Schwimmsandvorkommen durch geoelektrische Verfahren. I. Teil. Braunkohle **44**, Nr. 1/2, 6—9 (1945). — Einiges über den Blitzschutz besonders gefährdeter Objekte. Z. ges. Schieß- u. Sprengstoffwes. **39**, Nr. 7, 8, 9, 101—108 (1944). — Meßverfahren der Funkmutung. München: Verlag Oldenbourg. 1943.

(Eingegangen am 7. März 1946.)

Grundzüge einer Algebra der elektrischen Schaltungen.

Von **O. Plechl** und **A. Duschek**, Wien.

Mit 29 Textabbildungen.

Übersicht.

Die vorliegende Arbeit zeigt die Möglichkeit einer rechnerischen Behandlung der elektrischen Schaltungen, die allerdings erst die Entwicklung eines im wesentlichen völlig neuen algebraischen Kalküls erfordert. Nach einer Definition der Grundbegriffe in 1. werden die Grundlagen des Kalküls dargestellt und an Beispielen erläutert. Zum Schluß (11.) wird noch auf den Zusammenhang mit dem Aussagenkalkül der algebraischen Logik eingegangen.

Elektrische Schaltungen sind die Verbindungsglieder zwischen Stromquellen und Verbrauchern, worunter wir alle elektrischen Maschinen und Apparate verstehen, in welchen eine Energieumwandlung stattfindet. Die Aufgabe der Schaltung ist es, gewisse Gruppen von Verbrauchern an gewisse Gruppen von Stromquellen anzuschließen oder sie von diesen abzuschalten, d. h. also, gewisse Strompfade herzustellen oder zu unterbrechen. Die Strompfade müssen vorgeschrieben sein, mitunter ist auch vorgeschrieben, daß die Herstellung oder Lösung der Strompfade von bestimmten verschiedenen Stellen aus zu erfolgen hat. Ein einfaches Beispiel hierfür ist die Wechselschaltung, bei der ein Verbraucher (Glühlampe) von zwei oder mehr verschiedenen Stellen aus ein- oder ausgeschaltet werden kann. Aus diesen Schaltbedingungen, die den gegebenen Ausgangspunkt bilden, hat dann der Konstrukteur die Schaltung selbst, d. h. ihre graphische Darstellung, das Schaltbild zu ermitteln. Der Weg dazu bestand bisher ausschließlich im Probieren, wobei natürlich dem Konstrukteur seine mehr oder minder große praktische Erfahrung, seine Routine, zu Hilfe kommt. Irgendeine Systematik, ja auch nur der Ansatz zu einer solchen fehlte bisher vollständig, wenn man von den Ansätzen von Edler[1] und Lischke[2] absieht, und daher war man

[1] R. Edler: Der Entwurf von Schaltungen und Schaltapparaten (Schaltungstheorie), 2 Bände. Leipzig. 1905 und 1927.

[2] Lischke: Schaltlehre. Leipzig. 1921.

14*

A reference to the work by Robert Edler

auch nicht in der Lage zu erkennen, ob eine einmal gefundene Lösung auch die einfachste und zweckmäßigste sei. Meist ist man ja froh, überhaupt eine Lösung der manchmal recht komplizierten Schaltbedingungen zu haben.

Der Gedanke einer mathematischen Behandlung der elektrischen Schaltungen mag fürs erste recht befremdend wirken, da die Möglichkeit eines allgemeinen Ansatzes nicht ohne weiteres erkennbar ist. Nun gibt es zwar ein Gebiet der Mathematik, das sich in eine sehr enge Beziehung zu den Schaltungen bringen läßt, und zwar ist das der Aussagenkalkül der algebraischen Logik. Aber einerseits ist dieser Kalkül selbst vielen Mathematikern und erst recht den Praktikern kaum vom Hörensagen bekannt, anderseits geht das Interesse des Logistikers in einer ganz anderen Richtung als das unsere, so daß der Aussagenkalkül, dessen ganzer Ausbau derzeit überhaupt noch etwas dürftig erscheint, nur wenig mehr als eben die bloße Erkenntnis bringt, daß eine mathematische Theorie der elektrischen Schaltungen überhaupt möglich ist und daß es sich dabei um eine Art Algebra handelt. Wir werden im folgenden diese Algebra der elektrischen Schaltungen unabhängig vom Aussagenkalkül entwickeln und auf den Zusammenhang mit diesem erst zum Schluß kurz eingehen.

Die Hauptschwierigkeit jeder Anwendung der Mathematik liegt im Übergang von dem konkreten, in Worten formulierten physikalischen oder technischen Problem zu seiner mathematischen Fassung, also im mathematischen Ansatz. Dasselbe gilt natürlich auch von der Schaltalgebra, obwohl hier bei gewissen Schaltbedingungen, wenn die Zahl der Schalter vorgeschrieben ist, der mathematische Ansatz ungemein einfach ist. Auf der anderen Seite aber sind dem Schaltungstechniker die einfacheren, mitunter aber auch kompliziertere Schaltungen so geläufig, daß ihm in solchen Fällen die mathematische Behandlung keine Vorteile mehr geben kann. Wir glauben aber doch, daß es verfehlt wäre, die ganze Schaltalgebra von vornherein aus diesem Grunde als überflüssig abzulehnen. Abgesehen davon, daß schon die Möglichkeit einer solchen Disziplin an sich von größtem prinzipiellen Interesse sein dürfte, besteht gar kein Zweifel, daß der Kalkül, wenn man sich einmal mit ihm richtig vertraut gemacht hat, eine wertvolle Bereicherung des mathematischen Rüstzeuges des Elektrotechnikers darstellt und zu durchaus neuen, praktisch wichtigen Gesichtspunkten führt.

Die leitende Idee einer mathematischen Behandlung der Schaltungen, die Erkenntnis ihrer Möglichkeit und die Entwicklung des Kalküls, zunächst für eigene, rein praktische Zwecke stammt ausschließlich von Dr. O. Plechl. Er hat bemerkt, daß man sehr schön mit den Kontakten rechnen kann, wenn man sie mit irgendwelchen Buchstaben a, b, ... bezeichnet und dann die Parallelschaltung durch $a + b$, die Serienschaltung durch $a \cdot b$ wiedergibt. Etwas ähnliches machen auch Nakasima und Hanzawa,[3] die aber die Parallelschaltung mit $a \cdot b$ und die Serienschaltung mit $a + b$ bezeichnen. Das ist natürlich ganz unwesentlich und bedeutet von unserem Standpunkt aus nur, daß an Stelle der Leitwerte die Widerstandswerte der Rechnung zugrunde gelegt werden. Plechl hat weiter die formalen Gesetze der auf diesen beiden Grundoperationen aufgebauten Algebra gefunden und den Kalkül auf die verschiedensten Probleme der Schaltungstechnik angewendet. Die Begriffe Grenzfunktion, Vollform und Kürzung, all das stammt von Plechl, über dessen Gedankengang und Rechenmethoden in mancher Hinsicht die Arbeiten von H. Piesch[4, 5] besseren Aufschluß geben als die vorliegende, in der der Bearbeiter vor allem durch die grundsätzliche Einführung des Leitwertes einer Admittanz dem ganzen Kalkül eine ein-

[3] A. Nakasima und M. Hanzawa: The theory of equivalent transformation of simple partial paths in the relay circuit. Nippon electr. Commun. Engng. **9**, 32 (1938).

[4] H. Piesch: Begriff der allgemeinen Schaltungstechnik. Arch. Elektrotechn. **33**, 672 (1937).

[5] H. Piesch: Über die Vereinfachung von allgemeinen Schaltungen. Arch. Elektrotechn. **33**, 672 (1937).

Robert Edler

The book by R. Edler has been very influential on the research in this area in Central Europe.

Schaltlehre

Entwicklung und Anwendungen

Von

Dr. techn. Diplom-Ingenieur

Robert Edler

emer. a. o. Professor und Ehrenbürger der Technischen Hochschule in Wien
Ehrenmitglied des Elektrotechnischen Vereines Österreichs

Dritte, vollständig umgearbeitete Auflage

Mit 140 Abbildungen im Text

Wien
Franz Deuticke
1952

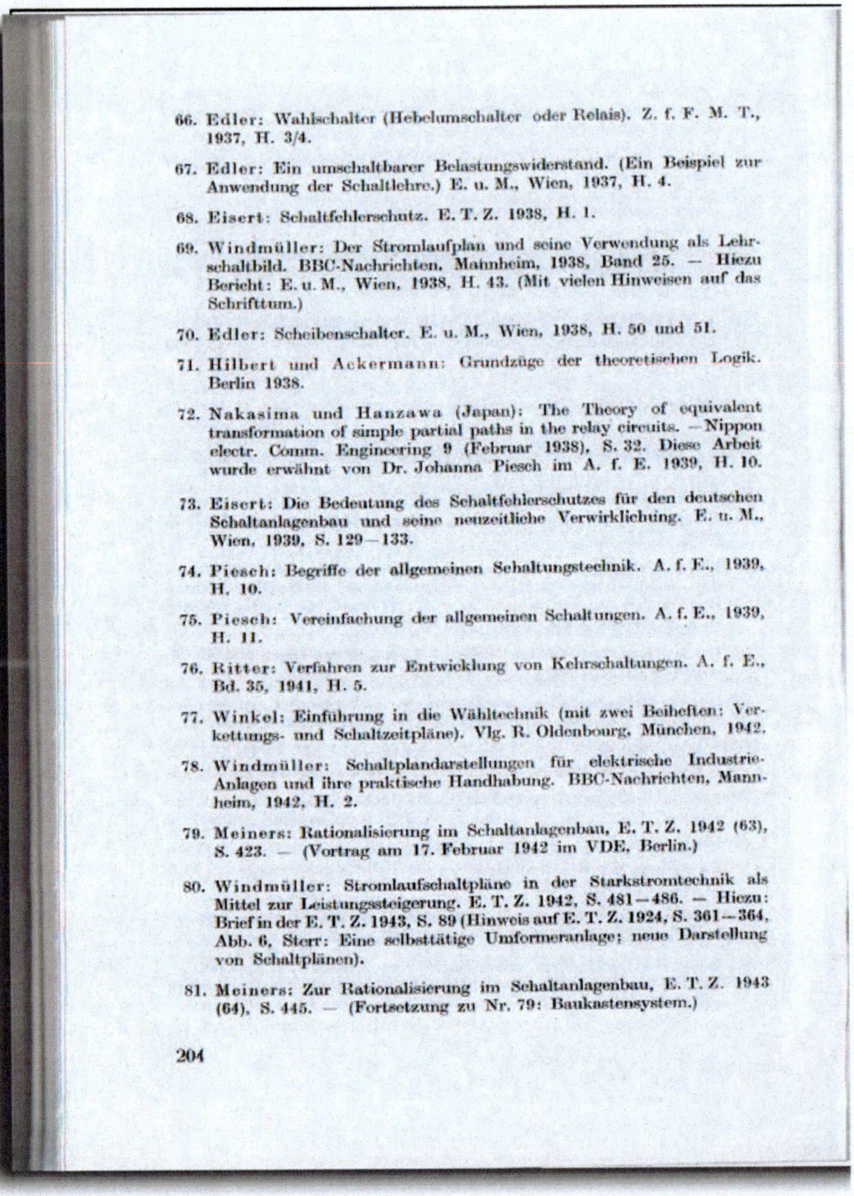

66. Edler: Wahlschalter (Hebelumschalter oder Relais). Z. f. F. M. T., 1937, H. 3/4.

67. Edler: Ein umschaltbarer Belastungswiderstand. (Ein Beispiel zur Anwendung der Schaltlehre.) E. u. M., Wien, 1937, H. 4.

68. Eisert: Schaltfehlerschutz. E. T. Z. 1938, H. 1.

69. Windmüller: Der Stromlaufplan und seine Verwendung als Lehrschaltbild. BBC-Nachrichten. Mannheim, 1938, Band 25. — Hiezu Bericht: E. u. M., Wien, 1938, H. 43. (Mit vielen Hinweisen auf das Schrifttum.)

70. Edler: Scheibenschalter. E. u. M., Wien, 1938, H. 50 und 51.

71. Hilbert und Ackermann: Grundzüge der theoretischen Logik. Berlin 1938.

72. Nakasima und Hanzawa (Japan): The Theory of equivalent transformation of simple partial paths in the relay circuits. — Nippon electr. Comm. Engineering 9 (Februar 1938), S. 32. Diese Arbeit wurde erwähnt von Dr. Johanna Piesch im A. f. E. 1939, H. 10.

73. Eisert: Die Bedeutung des Schaltfehlerschutzes für den deutschen Schaltanlagenbau und seine neuzeitliche Verwirklichung. E. u. M., Wien, 1939, S. 129—133.

74. Piesch: Begriffe der allgemeinen Schaltungstechnik. A. f. E., 1939, H. 10.

75. Piesch: Vereinfachung der allgemeinen Schaltungen. A. f. E., 1939, H. 11.

76. Ritter: Verfahren zur Entwicklung von Kehrschaltungen. A. f. E., Bd. 35, 1941, H. 5.

77. Winkel: Einführung in die Wähltechnik (mit zwei Beiheften: Verkettungs- und Schaltzeitpläne). Vlg. R. Oldenbourg, München, 1942.

78. Windmüller: Schaltplandarstellungen für elektrische Industrie-Anlagen und ihre praktische Handhabung. BBC-Nachrichten, Mannheim, 1942, H. 2.

79. Meiners: Rationalisierung im Schaltanlagenbau, E. T. Z. 1942 (63), S. 423. — (Vortrag am 17. Februar 1942 im VDE, Berlin.)

80. Windmüller: Stromlaufschaltpläne in der Starkstromtechnik als Mittel zur Leistungssteigerung. E. T. Z. 1942, S. 481—486. — Hiezu: Brief in der E. T. Z. 1943, S. 89 (Hinweis auf E. T. Z. 1924, S. 361—364, Abb. 6, Sterr: Eine selbsttätige Umformeranlage; neue Darstellung von Schaltplänen).

81. Meiners: Zur Rationalisierung im Schaltanlagenbau, E. T. Z. 1943 (64), S. 445. — (Fortsetzung zu Nr. 79: Baukastensystem.)

204

Reference to the work by Nakashima and Hanzawa, Piesch, Pleschl, Pleshcl and Duschek, and Shannon.

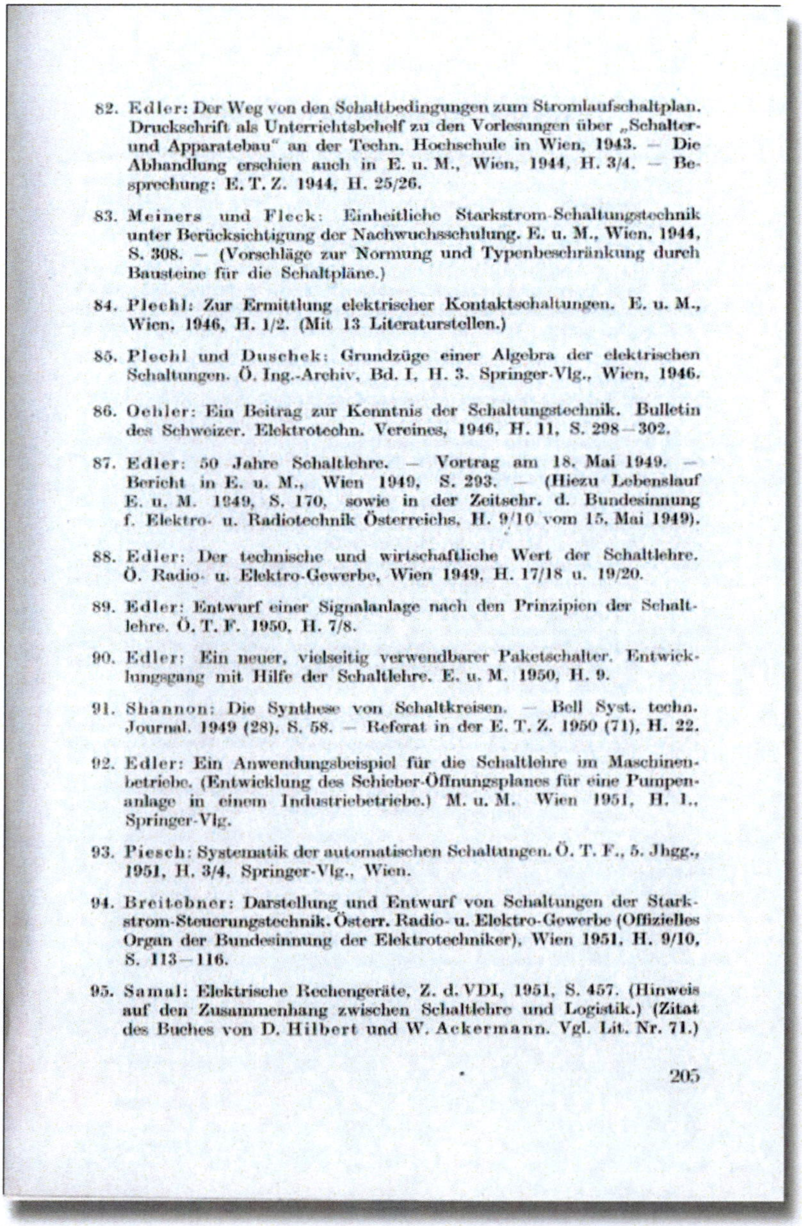

82. Edler: Der Weg von den Schaltbedingungen zum Stromlaufschaltplan. Druckschrift als Unterrichtsbehelf zu den Vorlesungen über „Schalter- und Apparatebau" an der Techn. Hochschule in Wien, 1943. — Die Abhandlung erschien auch in E. u. M., Wien, 1944, H. 3/4. — Besprechung: E. T. Z. 1944, H. 25/26.

83. Meiners und Fleck: Einheitliche Starkstrom-Schaltungstechnik unter Berücksichtigung der Nachwuchsschulung. E. u. M., Wien, 1944, S. 308. — (Vorschläge zur Normung und Typenbeschränkung durch Bausteine für die Schaltpläne.)

84. Plechl: Zur Ermittlung elektrischer Kontaktschaltungen. E. u. M., Wien, 1946, H. 1/2. (Mit 13 Literaturstellen.)

85. Plechl und Duschek: Grundzüge einer Algebra der elektrischen Schaltungen. Ö. Ing.-Archiv, Bd. I, H. 3. Springer-Vlg., Wien, 1946.

86. Oehler: Ein Beitrag zur Kenntnis der Schaltungstechnik. Bulletin des Schweizer. Elektrotechn. Vereines, 1946, H. 11, S. 298—302.

87. Edler: 50 Jahre Schaltlehre. — Vortrag am 18. Mai 1949. — Bericht in E. u. M., Wien 1949, S. 293. — (Hiezu Lebenslauf E. u. M. 1949, S. 170, sowie in der Zeitschr. d. Bundesinnung f. Elektro- u. Radiotechnik Österreichs, H. 9/10 vom 15. Mai 1949).

88. Edler: Der technische und wirtschaftliche Wert der Schaltlehre. Ö. Radio- u. Elektro-Gewerbe, Wien 1949, H. 17/18 u. 19/20.

89. Edler: Entwurf einer Signalanlage nach den Prinzipien der Schaltlehre. Ö. T. F. 1950, H. 7/8.

90. Edler: Ein neuer, vielseitig verwendbarer Paketschalter. Entwicklungsgang mit Hilfe der Schaltlehre. E. u. M. 1950, H. 9.

91. Shannon: Die Synthese von Schaltkreisen. — Bell Syst. techn. Journal. 1949 (28), S. 58. — Referat in der E. T. Z. 1950 (71), H. 22.

92. Edler: Ein Anwendungsbeispiel für die Schaltlehre im Maschinenbetriebe. (Entwicklung des Schieber-Öffnungsplanes für eine Pumpenanlage in einem Industriebetriebe.) M. u. M. Wien 1951, H. 1., Springer-Vlg.

93. Piesch: Systematik der automatischen Schaltungen. Ö. T. F., 5. Jhgg., 1951, H. 3/4, Springer-Vlg., Wien.

94. Breitebner: Darstellung und Entwurf von Schaltungen der Starkstrom-Steuerungstechnik. Österr. Radio- u. Elektro-Gewerbe (Offizielles Organ der Bundesinnung der Elektrotechniker), Wien 1951, H. 9/10, S. 113—116.

95. Samal: Elektrische Rechengeräte, Z. d. VDI, 1951, S. 457. (Hinweis auf den Zusammenhang zwischen Schaltlehre und Logistik.) (Zitat des Buches von D. Hilbert und W. Ackermann. Vgl. Lit. Nr. 71.)

205

Reference to the work by Nakashima and Hanzawa, Piesch, Pleschl, Pleshcl and Duschek, and Shannon.

Chapter 10
Postscript

Engineering in general and, thus, also computing, are supposed to be entirely pragmatic fields with a vast variety of ways of solving concrete practical tasks. However, to be efficient, they have to be based on a solid theoretical basis. These theories typically develop via two somewhat different routes.

1. Many particular solutions, or "pieces of theory" have been accumulated, and at a certain point it is possible to connect them and thus create a unified theory.
2. The complexity of an engineering system that is being managed manually grows too large to be manageable any longer and it becomes necessary to build a formalism and a theoretical framework.

For instance, the famous Maxwell equations can be viewed as an example for the first case, while the development of the switching theory falls more into the second category. In the following, we will take a closer look at this development which has been presented in a pictorial way in this book.

Digital systems at a hardware level basically consist of circuits and networks based upon two stable states. Mathematical models describing the functionality of these basic components are closely related to binary (Aristotelian) logic and mathematical logic. At higher functional levels, digital systems exhibit features that are related to number theory, graph theory, and discrete mathematics in general. Then, it is natural that many parts of these (and other) fields have been tied together to form the theoretical foundations of the design and analysis of digital systems and collectively constitute the Switching theory.

In this book, we present some historical observations about the early development of switching theory, especially the application of Boolean algebra. We want to draw attention to the contributions of scholars from different parts of the world, many of which cannot be found in the standard literature.

10.1 Switching Theory

Switching theory, and its more practical form logic design, provide mathematical foundations and tools for digital system design, and thus are essential in almost all areas of modern technology.

R.S. Stanković & J. Astola: From Boolean Log. to Switch. Circuits & Automata, SCI 335, pp. 183–194.
springerlink.com

The Boolean algebra occupies a central role in switching theory, and was the vehicle needed to transfer circuit design from art and skills devised from experience into a scientific discipline. Therefore, its basic concepts will be briefly presented in the following section.

10.1.1 Boolean Algebra

The Algebra of logic, also called the *symbolic logic*, is a method of expressing logic in a mathematical context. Instead of dealing with numeric quantities as in arithmetic, it is used to represent the truth value of a statement by assigning logic symbols 0 and 1 to two possible truth values *false* and *true*. It was derived by George J. Boole in order to permit an algebraic manipulation of logic statements and, therefore, is often called the Boolean algebra. It is useful in the study of information theory, set theory, probability theory, and as it has already been stated, represents the basis of switching theory.

The Boolean algebra can be defined as follows.

Definition 2 *(Boolean algebra)*
Consider a set B of at least two distinct elements 0 *and* 1. *Assume that there are defined two binary operations* $\vee$ *and* $\cdot$, *and the unary operation on B, usually called logic disjunction (OR), logic conjunction (AND) and logic negation (NOT), respectively.*

An algebraic system $(B, V, \cdot, 0, 1)$ *is a Boolean algebra if for any* $a, b, c \in B$ *the axioms in Table 10.1 are satisfied.*

Table 10.1 Axioms and postulates in Boolean algebra

1.	Idempotence	$a \vee a = 1,\ a \cdot 1 = a$
2.	Commutativity	$a \vee b = b \vee a,\ a \cdot b = b \cdot a$
3.	Associativity	$a \vee (b \vee c) = (a \vee b) \vee c$
		$a \cdot (b \cdot c) = (a \cdot b) \cdot c$
4.	Absorption	$a \vee (a \cdot b) = a,\ a \cdot (a \vee b) = a$
5.	Distributivity	$a \vee (b \cdot c) = (a \vee b) \cdot (a \vee c)$
		$a \cdot (b \vee c) = (a \cdot b) \vee (a \cdot c)$
6.	Involutivity	$\overline{\overline{a}} = a$
7.	Complement	$a \vee \overline{a} = 1,\ a \cdot \overline{a} = 0$
8.	Identity	$a \vee 0 = a,\ a \cdot 1 = a$
9.		$a \vee 1 = 1,\ a \cdot 0 = 0$
10.	DeMorgan Laws	$\overline{(a \vee b)} = \overline{a} \cdot \overline{b},\ \overline{a \cdot b} = \overline{a} \vee \overline{b}$

The postulates 2, 5, 7, and 8 are called the Huntington postulates [37], and are sufficient to specify a Boolean algebra, since the remaining six postulates can be derived from them.

10.2 Algebra of Logic

In this section, we discuss the development of the theoretical foundations of switching theory.

10.2.1 Work by G. J. Boole

In the spring of 1847, George J. Boole wrote a pamphlet entitled "Mathematical Analysis of Logic" [11], which he himself soon after regarded as an imperfect exposition of his logical system, and always rather referred directly to the later much elaborated treatise [13] as the proper expression of his considerations.

Boole did not view logic as a branch of mathematics, but emphasized an analogy between the symbols of algebra and the symbols that can be used to represent logical forms and syllogisms. By unity Boole meant the universe of thinkable objects, while literal symbols were used with elective meaning attached to common adjectives and substantives. With the use of such symbols, deriving syllogistic conclusion can be expressed in the form of equations. Manipulation with logic expressions should be used to

1. Demonstrate the truth value of a statement,

2. Rephrase a complicated statement in a simpler more convenient form without changing its meaning.

The latter feature is actually the foundation for engineering applications of the Boolean algebra.

In [13], Boole also discussed probability theory, which was not considered in [11], however, their logic parts are mainly the same.

The underlying idea of his work was to reduce the logical thought to the solving of equations. The algebraic operations have been defined to correspond to the basic activities during reasoning.

In terms of these operations Boole formulated in [13] an algebraic structure that shares the essential properties of both set operations and logic operations. The work was done independently of other works by logicians and mathematicians at that time. For instance, the results of Augustus De Morgan were not used, since Boole did not consider conjunction and disjunction a pair of dual operations.

It should, however, be noticed that Boole and De Morgan have had an extensive communication over a number of years [88].

The approach taken up by Boole and his point of view on the subject is possibly best described by himself. In [11], Boole wrote

That to the existing forms of Analysis a quantitative interpretation is assigned it is the result of the circumstances by which those forms were determined and is not to be construed into a universal condition of Analysis. It is upon the foundation of this general principle, that I purpose to establish the Calculus of Logic, and that I claim for it a place among the acknowledged forms of Mathematical Analysis.

In [12], Boole wrote as the introductory statements

In a work lately published, (meaning [11]) I have exhibited the application of a new and peculiar form of mathematics to the expression of the operations of the mind in reasoning. In the present essay I design to offer such an account of a portion of this treatise as may furnish a correct view of the nature of the system developed.

In [13], Boole wrote

The design of the following treatise is to investigate the fundamental laws of those operations of the mind by which reasoning is performed; to give expression to them in the symbolical language of a Calculus, and upon this foundation to establish the science of Logic and construct its method; and to make that method itself the basis of a general method for the application of the mathematical doctrine of Probabilities; and, finally, to collect from the various elements of truth brought to view in the course of these inquiries some probable intimations concerning the nature and constitution of the human mind.

As quoted in [15], it has been pointed out in [35] that Boole had planned to write a non-mathematical sequel to *The Laws of Thought* in which, as Boole said, he *would bring into light and prominence the philosophical elements which in my former exposition were too much hidden beneath the veil of symbolic notation*. The analysis in [35] was based on the study of some previously unpublished manuscripts treasured in the Royal Society. It has been pointed out that these manuscripts by Boole include *lengthy quotations from material apparently intended for this work*. As stated in [35], see also [15], most of these quotations pertain the analysis Boole did of the *faculty of conception* and to his development of the notion that *all true axioms* are somehow founded upon the *laws of the human mind*. From these comments, the author in [35] concluded that Boole had planned to include a non-algebraic expression of his logic, but this is not to be found among the present manuscripts.

Considerable contributions to formal formulation of Boolean algebra are due to William S. Jevons [40], [41], [42], [43], [44],[45], and Charles Sanders Pierce [66], [67]. See, also [14], [50]. In particular, Jevons exploited Boolean algebra in constructing a mechanical reasoning machine that was demonstrated to the Royal Society in 1870. For more details see, for instance [26].

Recall that in [97], Venn introduced a way for diagramming notation by Boole, which is now called the *Venn diagrams*.

It can be noticed that the method of visualization of logical relationships by Venn, has been elaborated on by Charles Lutwidge Dodson, more widely known as Carroll Lewis, the author of famous children novels, including for example *Alice in Wonderland* and *Through the Looking Glass*.

A systematic presentation of Boolean algebra and distributive lattices was given in 1890 by Ernst Schröder [81].

It should be noticed that Ernst Schröder developed his algebraic logic, which is nowadays called symbolic logic, independently of the work by G.J. Boole and A. De Morgan [25], about whose work he learned in 1873. Schröder published his work in [81], the third volume that appeared posthumously was edited by Eugen Möller.

In [80], Schröder presented Boolean algebraic logic ideas and in this ways supported the spreading of these ideas to continental Europe. See, also [36].

Boolean algebra presented as an axiomatic algebraic structure is due to Edward Vermilye Huntington in 1904 [37], [38], [39]. Marshall Harvey Stone in 1936 [93], and Garrett Birkhoff in 1940 [9], and in 1941 Birkhoff and Sanders MacLane [10] raised Boolean algebra to the level of deep mathematical subjects.

10.2.2 Work by Platon S. Poreckij

As it was pointed out in [5], Platon Sergeevic Poreckij was motivated to study logic by the famous mathematician A.V. Vasilev, and father of the founder of imaginary logic N.A. Vasliev, see [91].

Poreckij was primarily interested in logic equations and inequalities, and applications of mathematical logic in probability theory. As noticed in [4], the methods developed by Poreckij in this area were more universal than the approaches of Jevons and Venn [97], [98], at least as it has been estimated by Couturat [23].

In [74], is the first attempt at a complete theory of qualitative inference, where by the term quality Poreckij meant one-place predicate in modern terminology [1], see also [47]. In [75], Poreckij described his work by writing the following comment as a footnote on the first page.

In 1884, I published an article "On methods for solving logic equations", where it has been presented a complete theory of these equations. In this article, I suggest to exploit this theory in solving the following task in the Probability Theory.

Determine probability of a complex event, depending on given simple events, by using probabilities of these simple events as well as probabilities of some other complex events, assuming that given events satisfy an arbitrary number of arbitrary conditions.

A solution for this task was provided by Boole in his article [13], which, however, can hardly be considered scientific, since it is based upon arbitrary and entirely empirical theory of logic, in addition to the fact that the idea itself about the transition from logic equation to algebraic equations has weakly been elaborated by Boole. In this way, the main goal of the present paper is to give a scientific form to the deep, but vague and without proof, idea of Boole about the applicability of Mathematical Logic in Probability Theory.

A detailed information about Poreckij, and his work, including also a biography, can be found in [4], [5], [6], [27], and [92].

10.3 Applications of Boolean Algebra

In this section, we will discuss the history of the applications of logic, and in particular Boolean algebra, in circuit design, and in this way, the origins of the switching theory.

10.3.1 First Applications of Algebra of Logic in Circuit Design

In [102], and [103], it is reported that in a paper presented at the *Seminar on Logic* at the *Institute of Logic of the Academy of Science of USSR*, Geleii Nikolaevič Povarov discussed combinatory logic, distinguishing technical logic from the algebra of logic and pointing out that the possibility of technical applications of mathematical logic (i.e. the application of Boolean algebra to the analysis of electric relay-contact circuits) was first noted by the Russian physicist P.S. Ehrenfest in 1910 in a review of the work by Louis Couturat [23]. See also [22]. The same point of view has been supported also in [24]. For the contribution of Povarov to the research work on this subject, see for instance [76], and for more references of him [46] and [90].

The Russian edition of the book by Couturat was also reviewed in 1913 by Parentsev [64].

10.3.2 Work by V. I. Šestakov

As reported in [1], see also [19], [47], this proposal by Ehrenfest was elaborated in detail by Viktor Ivanovič Šestakov in 1934 and 1935, and was published in 1941.

In [1], it has been pointed out that the ideas Ehrnfest suggested about an algebra of switching circuits have been explored and elaborated by V.I. Šestakov, a student of V.I. Glivenko [32], [33], and the results were reported in written form in January 1935, however, Anovskaja states that this paper was not published at the time, and instead was used as the foundations for the PhD candidate Thesis written by Šestakov [83]. See also [46].

The major part of the thesis was published in the journal *Techničeska fizika*, Vol. 11, No. 6, 1941, [84], and also in [84], with further considerations in [86], [87]. As it has been pointed out in [46], credit to Šestakov is often given by many Soviet authors, as for example, in [65], although the work of Shannon is also well recognized and referred to. See, also [3], [5], [6].

The problem was discussed further by Mikhail Aleksandrovič Gavrilov in a series of papers published between 1943 [29] and 1947. The book by Gavrilov [30] established a basis for further study of switching theory in the Soviet Union, and since it was translated in German, had considerable influence abroad. Information on the publications on this subject by Gavrilov, Povarov, Šestakov and their associates can be found in [46], [48], and [90].

As noticed in [46], related work by V.A. Rozenberg, [77], [78], and M.G. Tsibalistii [95], has not been especially advantageously reported. The work done by Šestakov, Gavrilov and their followers has been reviewed in [49].

10.3.3 Work by C. E. Shannon

A revolutionary contribution to logic circuit design was made by Clode Elwood Shannon in his MSc. Thesis in 1938 at MIT. Shannon observed that Boolean algebra provides a rather complete mathematical foundation for relay circuit design and established, in his terminology, *a direct analogue between the calculus of propositional logic and symbolic relay analysis.*

In [82], Shannon used the following correspondence between the circuits and logic symbols.

1. If a terminal is open, it has an infinite impedance and the logic value 1 is assigned to it.

2. For a closed terminal, the impedance is zero, and the logic value 0 is correspondingly assigned.

3. Negation X for a terminal X is defined as the value opposite to the value assigned to X.

The variable X_{ab} connecting two terminals a and b is viewed as a function of time and called hindrance of the two terminal circuit ab. For the sake of simplicity, the symbol X_{ab} is often replaced by X whenever this does not cause any ambiguity.

For the simple relay locking circuit, Shannon uses the relation $X = A(B+X)$ and interprets it as the relay X operates iff either circuit A is closed or both circuit B is closed and relay X is operating. In a review of the work by Shannon, Alonzo Church [21] suggests using the notation as $X(t+1) = A(t) \vee (B(t)AX(t))$ where the time t is explicitly taken as an argument. The analysis and synthesis method for circuits has been expressed in the following way [82]

Any circuit is represented by a set of equations, the terms of the equations corresponding to the various relays and switches in the circuit. A calculus is developed for manipulating these equations by simple mathematical processes, most of which are similar to ordinary algebraic algorithms.

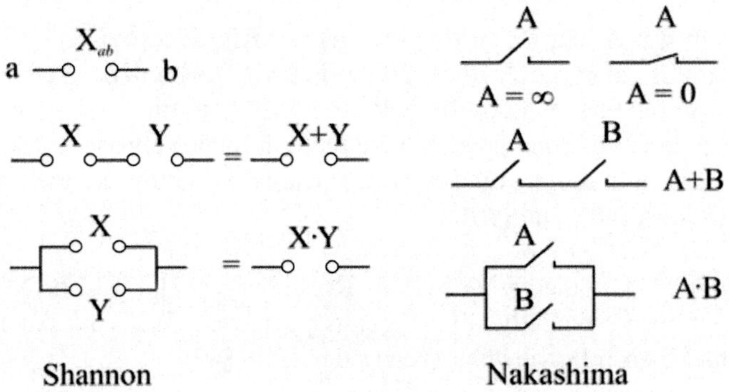

Fig. 10.1 Notation used by Shannon and Nakashima

For the synthesis problem the desired characteristics are first written as a system of equations, and the equations are then manipulated into the form representing the simplest circuit.

Fig. 10.1 illustrates the notation and symbols used by Shannon and Nakashima discussed above.

Example 1 *Fig. 10.2 shows an example of relay circuit realizing the function* $f(x,y,z,w) = w(x+y(z+\overline{x}))$ *in the Shannon notation and symbols.*

Shannon entered the University of Michigan in 1932, where he attended a course where he learned about Boolean algebra. Shannon was well informed about the large bibliography on symbolic logic [8][1], given as the first reference in his work [82], and in particular Shannon provided references to the English versions of [23], [100], and [38].

Shannon defended his MSc Thesis in 1938 and for this work was awarded in 1940 by the Alfred Noble American Institute of American Engineers Award given to a

[1] Notice that additions and corrections to this bibliography can be found in [17].

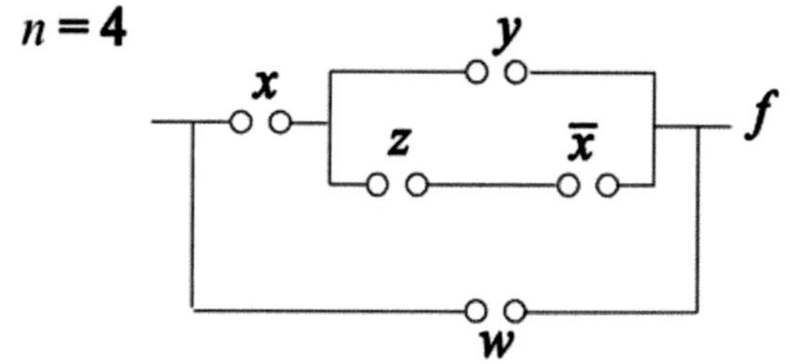

$$f = w(x + y(z + \bar{x}))$$

Fig. 10.2 Realization of the function f in Example 1

young author of a technical paper of exceptional merit, having been nominated for this award by his boss Vannevar Bush without his knowledge, in whose Lab Shannon wrote the thesis. For more details see, [16].

For the simplest introduction into the work by Shannon [82] the shortest way is to read the review by Charles A. Baylis in The *Journal of Symbolic Logic* [2] [2].

An interpretation in terms of relay and switching circuits of the two-valued Boolean algebra (the author calls it an analogue of the classical propositional calculus). The variables are circuits, $=0$ *means that the circuit is closed,* $=1$ *means the circuit is open,* $+$ *signifies a series connection of circuits, juxtaposition or* $\cdot$ *a parallel connection. Switches are so arranged that whenever a circuit x is open, there is a circuit x' which is closed, and whenever x is closed* x' *is open. Two circuits are equivalent when they are open and closed together. The author shows that every function of the algebra, thus interpreted, represents a seriesparallel circuit, and each such is represented by a function of the algebra. He shows further that certain types of non-series-parallel circuits are represented by and may be discovered from symmetric functions, and that equations of the algebra may be written to describe the operating characteristics of such circuits.*

[Review by C.A. Baylis]

10.3.4 Work by A. Nakashima

Akira Nakashima graduated from Tokyo University, and worked as an engineer at the *Nippon Electric Company* (NEC), working among other tasks, on the design of relay networks for various purposes.

[2] At the time when he was writing this review, C.A. Baylis was a Professor at the Department of Philosophy at Brown University, and was a founder of the Association of Symbolic Logic, and among the founding editors of The Journal of Symbolic Logic. He is a coauthor of [7].

Nakashima first did an extensive analysis of many case studies of relay networks trying to formulate a unified design theory for such networks. He considered impedances of relay contacts as two-valued variables, and used logic OR and AND operations to represent their series and parallel connections, respectively. Based on that, he formulated a related theory of relay networks by introducing and exploiting some algebraic relations that represent a basis of switching theory. For instance, he defined the rules that are nowadays called De Morgan duality expressions. These results, Nakashima presented without using a symbolic notation in a series of articles in the monthly journal *Nichiden Geppo*[3] of the Nipon Electric Company (NEC) entitled *Theory and Practice of Relay Engineering* [55]. The Telegraph and Telephone Society of Japan engaged Nakashima to give an invited talk at the annual meeting of this society early in 1935. This three-hours-long talk was later published in [51], [52].

In 1936, Nakashima was transferred to transmission engineering. However, having been advised to continue this research by Niwa Yasujiro, the Chief Engineer of NEC at that time, he continued the work after official office hours with the help of Masao Hanzawa, who remained on the exchange engineering team.

In a joint work with Masao Hanzawa, the theory of Nakashima was elaborated by the additional use of symbolic representations and finally evolved into an algebraic structure, about which Nakashima and Hanzawa concluded in August 1938 that it was actually equal to Boolean algebra.

Notice that the papers by Nakashima and also those coauthored with Hanzawa, were first published in Japanese in The Journal of the Institute of Telephone and Telegraph Engineers of Japan and The *Journal of the Institute of Electrical Communication Engineers*, and then later translated in reduced from and published in *Nippon Electrical Communication Engineering*.

In [55], which was published in Japanese in August 1937, the algebra introduced by Nakashima and elaborated on in cooperation with Hanzawa was reduced to an algebra of sets by assigning to each partial path a set of times at which its impedance is infinite. In that way, the author was able to state that "theorems and expressions developed in the theory of sets may, therefore, be applied to acting impedance problems of simple partial paths", see the corresponding remark in [19].

For instance, in [58] it is noticed the following correspondence between the algebra of logic and circuits. If A and B are two-terminal circuits, which are called *simple partial paths* by the authors, then $A+B$ and $A \cdot B$ correspond to their serial and parallel connections, respectively. The equation $A = B$ states that acting functions of A and B are equal, meaning that A is open (closed) when B is open (closed). Similarly, A denotes a simple partial path that is closed when A is open and vice versa. Two

[3] The current name of the journal is *NEC Technical Journal*.

simple partial paths that are always closed or open are denoted by p and s, respectively. In the terminology used by the authors, such paths have an infinite and zero impedance, respectively.

With this notation (see the corresponding page), that is in essence equal to those by Shannon and other authors, Nakashima and Hanzawa defined an algebra which, as they realized in 1938, is identical to the Boolean algebra. In particular, the authors noticed that their expansion theorem is actually the same one that Boole calls the *law of development.*

In [56], which was published in Japanese in August 1937, the author is pointing out the links between his algebra and the algebra of sets, which therefore can be used in discussing problems related to simple partial paths and the impedances in them [20]. See also [57].

In [60], Nakashima and Hanzawa for the first time refer explicitly to Boole and Schröder.

In [99], the work by Nakashima [53] was estimated to be the first systematic study of logic circuits. What was especially emphasized was the expansion theorem of impedance function [62] of relay circuits and the design theory of two-terminal relay circuits in 1940.

In [101], it is explicitly stated that Akira Nakashima published the first paper on switching theory in the world.

References to the work done by Nakashima are given by Shannon [82], and Piesch [69].

In Japan, the work done by Nakashima has been continued by Kan-ichi Ohashi, Mochinori Goto [34], Yasuo Komamiya, T. Kojima, and later by many others, see for example, [79], chapter entitled *Switching Theory in Japan.*

It is interesting to notice that Nakashima used the same interpretation of contacts in relays as Shannon did. The question about the opposite interpretation of open and closed contacts has been discussed in [63]. The answer provided by Nakashima and Hanzawa appeared in [61].

For more details about the work of Akira Nakashima, see [89].

10.3.5 Work by J. Piesch

A very considerable amount of work was carried out by the Austrian mathematician Johanna Piesch, whose name is often abbreviated as Hanna, or Hansi, [18], for different reasons.

In [68] and [69], we find circuits (switches) with any finite number of states (positions) and under the assumption that not all of them must have the same number of states. Switches are denoted by symbols, such as a, b, c, etc., with different positions indicated by subscripts. In this way, the symbol a_i will denote that the switch a is in the position i. Capital letters are used to express other propositions as effects caused by certain assignments of states, thus, to describe the outputs of the network.

The operations of addition, multiplication, and inverse, corresponding to disjunction, conjunction, and negation, are used to form expressions of the algebra built on propositions denoted by symbols. This algebraic structure is actually a propositional algebra, rather than propositional calculus, since only the equivalences are asserted [18].

In her work Johanna Piesch refers to [59] and an unpublished paper by the Austrian researcher Otto Pleschl, see for instance related later publications by the same author [71], [72]. In [72] she refers to [68], and [69], as well as [58].

In [101], it has been remarked that Piesch in publications dating from 1939, did an extension of the work by Nakashima.

It is interesting to point out that the method in [68], [69] has been used in [31] and adapted to the design of n-terminal switching circuits, see also [18].

Research work in this area in Austria has been done by Otto Pleschl and his associate A. Duschek [71], [72], [73]. The book by R. Edler has been very influential on other researchers in Central Europe.

References in the Postscript

[1] Anovskaja, S.A.: Osnovania matematiki i matematiceskaa logika (Foundations of mathematics and mathematical logic). In: Mathematika v SSSR za tridcat let 1917-1947 (Mathematics in the USSR for the thirty years 1917-1947), OGIZ, Moscow and Leningrad, pp. 9–50 (1948)

[2] Baylis, C.A.: Review of A Symbolic Analysis of Relay and Switching Circuits by Claude E. Shannon. The Journal of Symbolic Logic 4(2), 103 (1939)

[3] Bazanov, V.A.: V.I. Shestakov and C. Shannon - Different fates of one brilliant idea authors. Voprosy Istorii Estestvoznaniya i tehniki (2), 112–121 (2005) (in Russian)

[4] Bazanov, V.A.: Life and work of pioneer of mathematical logical studies in Russia. Voprosy Istorii Estetvoznaniya i tehniki (4), 64–73 (2005)

[5] Bazanov, V.A.: Ocerki sotsialnoj istorii logiki v Rossii (Essays in the Social History of Logic in Russia), Izdatelstvo Srednevolzhskogo nautchnogo tsentra (Ulyanovsk-Simbirsk, Mid-Volga Research Center Publ.) (2002)

[6] Bazanov, V.A.: History of Logic in Russia and the USSR - Conceptual Context of University Philosophy. Canon+, Moscow (2007)

[7] Bennett, A.A., Baylis, C.A.: Formal Logic - A Modern Introduction. Prentice-Hall, New York (1939)

[8] Bibliography of Symbolic Logic. The Journal of Symbolic Logic 1(4) (December 1936)

[9] Birkhoff, G.: Lattice theory,vol. 25, ch. 5&6. Amer. Math. Soc., Colloquium Publications (1940)

[10] Birkhoff, G., MacLane, S.: A Survey of Modern Algebra. Mcmillan Co., New York (1941)

[11] Boole, G.J.: Mathematical Analysis of Logic, being an essay towards a calculus of deductive reasoning, London, Cambridge, 82 pages (Spring 1847); Reprinted in Jourdain, P.E.B. (ed.): George Boole's Collected Logical Works, Chicago, London, vol. 1 (1916)

[12] Boole, G.J.: The calculus of logic. The Cambridge and Dublin Mathematical Journal 3, 183–198 (1848); Reprinted in Jourdain, P.E.B. (ed.): George Boole's Collected Logical Works, Chicago, London, vol. 1 (1916)

[13] Boole, G.J.: An Investigation of The Laws of Thought, on which are founded the mathematical theories of logic and probabilities, v+iv+424 pages (1854); reprinted in Jourdain, P.E.B. (ed.): George Boole's Collected Logical Works, Chicago, London, vol. 2 (1916); Reprinted by Dover Publications, Inc., New York (1954)

[14] Brown, F.M.: Boolean reasoning, The logic of Boolean equations. Kluwer Academic Publishers, Boston (1992)

[15] Chisholm, R.M.: Review of Boole's Philosophy of Logic by Mary B. Hesse. The Journal of Symbolic Logic 17(4), 285 (1952)

[16] Chiu, E., Lin, J., Mcferron, B., Petigara, N., Seshasai, S.: Mathematical Theory of Claude Shannon, Report 6.933J/STS.420J The Structure of Engineering Revolutions (2001)

[17] Church, A.: Additions and corrections to a bibliography of symbolic logic. The Journal of Symbolic Logic 3(4), 178–192 (1938)

[18] Church, A.: Archiv für Electrotechnik - review of papers. The Journal of Symbolic Logic 30(2), 247–248 (1952)

[19] Church, A.: Review of Formal logic and switching circuits by T.A. Kalin. In: Proceedings of the 1952 ACM national meeting (Pittsburgh), Pittsburgh, Pennsylvania, May 02, pp. 251–257 (1952); The Journal of Symbolic Logic 18(4), 345–346 (1953)

[20] Church, A.: Review of Nippon Electrical Communication Engineering, by Akira Nakasima and Masao Hanzawa. The Journal of Symbolic Logic 18(4), 346 (1953)

[21] Church, A.: Review of Synthesis of Electronic Computing and Control Circuits by Aiken, H.H., Burkhart, W., Kalin, T., Strong, P.F. The Journal of Symbolic Logic 18(4), 347 (1953)

[22] Comey, D.D., Kline, G.L.: Review of Rabote seminara po logiki v Institute Filosofii AN SSSR (Work of the Seminar on Logic in the Institute of Philosophy of the Academy of Sciences of the USSSR) by A. A. Zinoviev. The Journal of Symbolic Logic 24(3), 232–233 (1959)

[23] Couturat, L.: L algebre de la logique, Paris, 2nd edn., p. 100 pages (1905); Paris 100 pages (1914); Hungarian translation A logika algebraja, translated by Denes Konig. Mathematikai es physikai lapok, Budapest 17, 109–202 (1908); Russian translation Algebra logiki, Mathesis, Odessa, pp. iv+l07+xii+6 (1909); English version, The Open Court Publishing Company

[24] Durand-Richard, M.J.: Pour une historiographie comparee de l'histoire de l'informatique. Report Universite Paris 8-REHSEIS

[25] De Morgan, A.: Formal logic - or the calculus of inference, necessary and probable, London, xvi+336 (1847)

[26] Devlin, K.: Goodbye Descartes - The End of Logic and the Search for a New Cosmology of the Mind. John Wiley, Chichester (1997)

[27] Dubago, D.: P.S. Poreckij. Bull. Soc. Phys.-Math. de Kasan, 2s 16(1), 3–7 (1908)

[28] Ehrenfest, P.: Review of Couturat's Algebra logiki. Zurnal Russkago Fizikohemiceskago Obscestva, Fiziceskij otdel 42, 382–387 (1910)

[29] Gavrilov, M.A.: The synthesis and analysis of relay-contact networks. Avt. i Telem. 4 (1943)

[30] Gavrilov, M.A.: The Theory of Relay-Contact Networks, Izdat. Akad. Nauk SSSR, Moscow (1950); translation in German Relaisschalttechnik fur Stark-und Schwachstro-manlagen. V.E.B. Verlag Technik, Berlin (1963)

[31] Gilbert, E.N.: W-terminal switching circuits. The Bell System Technical Journal 30, 668–688 (1951)

[32] Glivenko, V.I.: Sur la logique de M. Brouwer. Acad. Belg., Bull. des. sci., 5s 14, 225–228 (1928)

[33] Glivenko, I.V.: Sur quelques points de la logique de M. Brouwer. Acad. Belg., Bull. des sci., 5s 15, 183–188 (1929)

[34] Goto, M.: Applications of logical equations to the theory of relay contact networks. Electric Soc. of Japan 69, 125 (1949)

[35] Hesse, M.: Boole's philosophy of logic. Annals of Science 8, 61–81 (1952)
[36] Houser, N.: Algebraic Logic from Boole to Schröder, 1840-1900. In: Grattan-Guinness Companion Encyclopedia of the History and Philosophy of the Mathematical Sciences. Routledge, London (1994)
[37] Huntington, E.V.: Sets of independent postulates for the algebra of logic. Trans. Amer. Math. Soc. 5(3), 288–309 (1904)
[38] Huntington, E.V.: New sets of independent postulates for the algebra of logic. Trans. Amer. Math. Soc. (35), 274–304 (1933)
[39] Huntington, E.V.: Boolean algebra: A correction. Trans. Amer. Math. Soc. 35, 557–558 (1933)
[40] Jevons, W.S.: Pure Logic or the Logic of Quality apart from Quantity with Remarks on Boole's System and the Relation of Logic and Mathematics. E. Stanford, London (1864)
[41] Jevons, W.S.: Pure Logic and Other Minor Works, Pure Logic or the Logic of Quality Apart From the Quantity. Macmillan and Co., London (1890)
[42] Jevons, W.S.: On the mechanical performance of logical inference. Philosophical Transactions of the Royal Society 160, 497–518 (1870)
[43] Jevons, W.S.: The Principles of Scientific Method, 2nd edn., 2 vols. Mcmillan & Co., London (1879)
[44] Jevons, W.S.: Studies in Deductive Logic - A Manual for Students. Mcmillan & Co., London (1880)
[45] Jevons, W.S.: Pure Logic and Other Minor Works. In: Adamson, R., Jevons, H.A. (eds.) Mcmillan & Co., London (1890); reprinted by Thoemmes Press, Bristol (1991)
[46] Kautz, W.H.: A survey and assessment of progress in switching theory and logical design in the Soviet Union. IEEE Ttans. Electronic Computers EC-15(2), 164–204 (1966)
[47] Kline, G.L.: Review of Foundations of Mathematics and Mathematical Logic by S.A. Anovskaa. The Journal of Symbolic Logic 16(1), 46–48 (1951)
[48] Kung, G.: Bibliography of soviet work in the field of mathematical logic and foundations of mathematics from 1917-1957. Notre Dame Journal of Formal Logic 3(1), 1–40 (1962)
[49] Ljapunov, A.A.: Matematiceskie isledovania sviazannye s eksploatacei elektronnukh vycislitelnih masin. Matematika v SSSR za sorok let, Moscow (1959)
[50] Mays, W., Henry, D.P.: Jevons and Logic. Mind - A Quarterly Review of Psychology and Philosophy LXII,, 448–505 (1953)
[51] Nakashima, A.: Synthesis theory of relay networks. Journal of the Telegraph and Telephone Society of Japan (September 1935)
[52] Nakashima, A.: Some properties of the group of single partial paths in the relay circuit. Journal of the Institute Telephone and Telegraph Engineers of Japan, 88–95 (February 1936); English translation in Nippon Electrical Communication Engineering, pp. 70–71 (March 1937)
[53] Nakashima, A.: Some properties of the group of simple partial paths in the relay circuit. Journal of the Institute of Telephone and Telegraph Engineers of Japan, 88–95 (January, February, March 1936); English translation in Nippon Electrical Communication Engineering, 70–71 (March 1937)
[54] Nakashima, A.: The theory of four-terminal passive networks in relay circuit. Journal of the Institute of Electrical Communication Engineers of Japan (April 1937); English summary in Nippon Electrical Communication Engineering (10), 178–179 (April 1938)

[55] Nakashima, A.: Algebraic expressions relative to simple partial paths in the relay circuits. Journal of the Institute of Electrical Communication Engineers of Japan (137) (August 1937) (in Japanese); Condensed English translation of about half the length of the original paper in Nippon Electrical Communication Engineering (12), 310–314 (September 1938); Section V, Solutions of acting impedance equations of simple partial paths
[56] Nakashima, A.: The theory of two-point impedance of passive networks in the relay circuit (Part 1). Journal of the Institute of Electrical Communication Engineers of Japan (177) (December 1937); Reduced version of part 1 and part 2 (bellow) appears in Nippon Electrical Communication Engineering (13), 405–412 (November 1938)
[57] Nakashima, A.: The transfer impedance of four-terminal passive networks in the relay circuit. Journal of the Institute of Electrical Communication Engineers of Japan (179) (February 1938); Condensed English version in Nippon Electrical Communication Engineering (14), 459–466 (December 1938)
[58] Nakashima, A., Hanzawa, M.: The theory of equivalent transformation of simple partial paths in the relay circuit (Part 2). Journal of the Institute of Electrical Communication Engineers of Japan (167) (December 14, 1936); published in February 1937; Condensed English version of parts 1 and 2 in Nippon Electrical Communication Engineering (9), 32–39 (February 1938)
[59] Nakashima, A.: Algebraic expressions relative to simple partial paths in the relay circuits. Journal of the Institute of Electrical Communication Engineers of Japan (173) (August 1937) (in Japanese); Condensed English translation of about half the length of the original paper in Nippon Electrical Communication Engineering (12), 310–314 (September 1938); Section V, Solutions of acting impedance equations of simple partial paths
[60] Nakashima, A., Hanzawa, M.: Expansion theorem and design of two-terminal relay networks (Part 1). Journal of the Institute of Electrical Communication Engineers of Japan (206) (May 1940); Condensed English version in Nippon Electrical Communication Engineering (24), 203–210 (April 1941)
[61] Nakashima, A., Hanzawa, M.: The answer to the above question and comments. Journal of the Institute of Electrical Communication Engineers of Japan, 665–666 (November 1941) (in Japanese)
[62] Nakashima, A., Hanzawa, M.: Expansion theorem and design of two-terminal relay networks (Part 2). Journal of the Institute of Electrical Communication Engineers of Japan (August 1940); Condensed English version in Nippon Electrical Communication Engineering (26), 53–57 (October 1941)
[63] Okada, S.: Question and comments about The law of development of impedance functions and theory of designing two-terminal networks in the relay circuit by Akira Nakashima and Masao Hanzawa. Journal of the Institute of Electrical Communication Engineers of Japan, 662–665 (November 1941) (in Japanese)
[64] Paretsev, N.N.: Neskol'ko slov po povodu knigi L. Kutura "Filosofskii principy matematiki" (A few words apropos of L. Couturat's book, "Filosofskie principy matematiki"). Izvestia Fiziko-Matematicaskago Obscestva pri Imepratorskom Kazanskom Universitete (Bulletin de la Society Physico-Mathématique de Kasan), ser. 2 19(1), 15–23 (1913)
[65] Petrov, V.V., Rutkhovskii, V.Y.: Some problems relative to the logical design of circuits and the selection of specifications for high-speed servomechanisms. Avt. Uprav. i Vych. Tekh. 2 (1959)
[66] Peirce, C. S.: On the algebra of logic. American Journal of Mathematics 3, 15–57 (1880)

[67] Peirce, C.S.: On the algebra of logic - A contribution to the philosophy of notation. American Journal of Mathematics 7, 180–203 (1884)
[68] Piesch, H.: Begriff der allgemeinen Schaltungstechnik. Archiv für Elektrotechnik 33, Heft 10, 672–686 (1939) (in German)
[69] Piesch, H.: Über die Vereinfachung von allgemeinen Schaltungen. Archiv fur Elektrotechnik 33, Heft 11, 733–746 (1939)
[70] Piesch, J.: Systematik der automatischen Schaltungen. O.T.F, 5. Jahrgang, Heft 3/4, pp. 29–43. Springer Verlag, Wien (1951)
[71] Plechl, O.: Zur Ermittlung elektrischer Kontaktschaltungen. E.u.M., Wien, H. 1/2, 34–38 (1946)
[72] Plechl, O., Duschek, A.: Grundzüge einer Algebra der elektrischen Schaltungen. In: O.Ing.-Archiv, pp. 203–230, Bd. I, H. 3. Springer Verlag, Wien (1946)
[73] Plechl, O.: With the help by Rieder, W., Elektromechanische Schaltungen und Schaltgerte - eine Einfhrung in Theorie und Berechnung, Issue Erg. u. bearb., Pulished by von Werner Rieder Erschienen, p 224. Springer, Wien (1956)
[74] Poreckij, P.S.: Referat on basic principles of mathematical logic. In: Protokol treti-ago zasedania sekcii fiziko-matematiceskih nauk Obscestva Estestvoispytatelej pri Imperatorskom Kazanskom Universitete, 17. maja 1880 goda, 2-30 (issued as a part of Protokoly zasedanij Obscestva Esytestvoispytelej pri Imperatorskom Kazanskom Universitete, Kazan, Russia, vol. 12 (1880-1881)
[75] Poreckij, P.S.: Solution of general tasks in probability theory through the mathematical logic. Izd-vo Kazanj Univ., Kazanj, Russia (1887); printed by the permission of the Society for Natural Sciences at the Imperial Üniversity of Kazan, the President of the Society A. Štukenberg
[76] Povarov, G.N.: Matrix methods of analyzing relay-contact networks in terms of the conditions of non-operation. Avt. i Telem. 15(4), 332–335 (1954)
[77] Rozenberg, V.A.: The problem of blocking and transformation of contact groups. Avt. i Telem. (1), 47–54 (1940)
[78] Rozenberg, V.A.: Some properties of relay selection. Avt. i Telem. (1), 37–48 (1939)
[79] Sasao, T.: Switching Theory for Logic Synthesis. Kluwer Academic Publishers, Dordrecht (1999); translation in Serbian Prekidačka teorija za logičko projektovanje by R.S. Stanković, Nauka, Belgrade (1999); 2nd edn., Nauka, Belgrade (2005)
[80] Schröder, E.: Der Operationskreis des Logik kalküls. B.G. Teubner, Leipzig (1877)
[81] Schröder, E.: Vorlesungen uber die Algebre der Logik (exacte Logik). B.G. Teubner, Leipzig (1890, 1891, 1895, 1905); reprinted in three volumes by Chelsea Publishing Company, Bronx, USA (1966); Thoemmes Press (2000)
[82] Shannon, C.E.: A symbolic analysis of relay and switching circuits. Transactions of the American Institute of Electrical Engineers 57, 713–723 (1938)
[83] Šestakov, V.I.: Some Mathematical Methods for Construction and Simplification of Two-element Electrical Networks of Class A, PhD Dissertation, Lomonosov State University, Moscow, Russia (1938)
[84] Šestakov, V.I.: The algebra of two-terminal networks constructed exclusively of two-terminal elements (The algebra of A-networks). Avtomatika i Telemekhanika 2(6), 15–24 (1941)
[85] Šestakov, V.I.: The algebra of two-terminal networks constructed exclusively of two-terminal elements (The algebra of A-networks). Žurnal Tekh. Fis. 11(6), 532–549 (1941)
[86] Šestakov, V.I.: A symbolic calculus applied to the theory of electrical relay networks. Uchenye Zapiski Moskowskog Gosudarstvenog Universiteta 73(5), 45–48 (1944)

[87] Šestakov, V.I.: The representation of the characteristic functions of propositions by means of expressions which are realized by relay-contact networks. Isv. Akad. Nauk., Ser. Matem. 10, 529–554 (1946)

[88] Smith, G.C.: The Boole - De Morgan Correspondence, pp. 1842–1864. Calderon Press, Oxford (1982)

[89] Stanković, R.S., Astola, J. (eds.): Reprints from the Early Days of Information Sciences, On the Contributions of Akira Nakashima to Switching Theory. TICSP Series # 40, Tampere International Center for Signal Processing, Tampere, Finland (2008) ISSN1456-2774, ISBN 978-952-15-1980-2

[90] Stanković, R.S., Sasao, T., Astola, J.T., Publications in the First Twenty Years of Switching Theory and Logic Design, TICSP Series #14, Tampere (2001), ISBN 952-15-0679-2

[91] Stelzner, W.: Review of Ocerki sotsialnoj istorii logiki v Rossii (Sketches of the Social History of Logic in Russia) by V. A. Bazhanov. The Bulletin of Symbolic Logic 10(3), 421–423 (2004)

[92] Stiazkin, N.I.: K voprosy o vklade P.S. Poreckog v razvitie matematiceskoi logiki (To the question of the contribution of P.S. Poreckij to the development of mathematical logic), pp. 103–109. Vestnik universiteta, ser. ekon. i prava, Moscow (1956)

[93] Stone, M.H.: The theory of representation for Boolean algebras. Trans. Amer. Math. Soc. 40, 37–111 (1936)

[94] Stone, M.H.: The representation of Boolean algebras. Bulletin of the American Mathematical Society 44, 807–816 (1938)

[95] Tsimbalistii, M.G.: The economical design of relay networks. Trudy Leningradskoi Eksperimentalnoi Elektrotekhnicheskoi Labor 8, 19–31 (1928)

[96] Tsimbalistii, M.G.: The problem of simplification in the design of relay networks. Trudy Leningradskoi Eksperimentalnoi Elektrotekhniceskoi Labor (8), 19–31 (1928)

[97] Venn, J.: On the diagrammatic and mechanical representation of propositions and reasonings. Dublin Philosophical Magazine and Journal of Science 9(59), 1–18 (1880)

[98] Venn, J.: Symbolic Logic. Macmillan, London, 446 pages (1881)

[99] Watanabe, H.: Activities on Circuit Theory in Japan. IEEE Trans. Circuits and Systems CAS-31, 112–123 (1984)

[100] Whitehead, A.N.: Universal Algebra, vol. 1, book 3, pp. 35–42. Cambridge University Press, Cambridge (1898)

[101] Yamada, A.: History of research on switching theory in Japan. IEE J. Trans. FM 124(8), 720–726 (2004)

[102] Zinoviev, A.A.: Raščirat tematiku logičeskih issledovanij (Broaden the subject matter of logical investigations). Voprosy filosofii (3), 211–215 (1957)

[103] Zinoviev, A.A.: O rabote seminara po logike v Institute Filosofii AN SSSR (Work of the Seminar of Logic in the Institute of Philosophy of the Academy of Sciences of the SSSR. Voprosy filosofii (2), 162–172 (1958)

Further Reading

References used in preparation of this booklet besides those mentioned in the Postscript.

1. Berstein, B.A.: A set of four independent postulates for Boolean algebras. Trans. Amer. Math. Soc. 17(1), 50–52 (1916)
2. Bibliography "Explanation of Bibliography". The Journal of Symbolic Logic 1(4), 121–123 (1936)
3. Bibliography "Items 1 (Leibniz) - 122 (Huntington)". The Journal of Symbolic Logic 1(4), 123-155 (December 1936)
4. Bibliography "Items 123 (Bocher) - 270 (Lenzen)". The Journal of Symbolic Logic 1(4), 155–183 (December 1936)
5. Bibliography "Items 271 (Mursell) - 547 (Barzin)". The Journal of Symbolic Logic 1(4), 183–216 (December 1936)
6. Bibliography "Index of Authors". The Journal of Symbolic Logic 1(4), 217–218 (December 1936)
7. Bibliography "Index of Subjects". The Journal of Symbolic Logic 1(4), 193–204 (December 1936)
8. Bibliography "Back Matter". The Journal of Symbolic Logic 1(4), 219 (December 1936)
9. Blachman, N.M.: Review of The foundations of computing machinery by John D. Goodell. The Journal of Symbolic Logic 4, 256 (1961)
10. Blachman, N.M.: The state of digital computer technology in Europe. Communications of the ACM 18(4), 348–349 (1953)
11. Black, M.: Review of Philosophical problems of many-valued logic by A.A. Zinoviev, Guido Kung, David Dismore Comey. The Journal of Symbolic Logic 75(3), 385–388 (1966)
12. Brusentsov, N.P., Vladimirova, Y.S.: Solution of Boolean equations. Computational Mathematics and Modeling 9(4), 287–295 (1988)
13. Carr, J.W.: A visit to computation centers in the Soviet Union. Communications of ACM 2, 8–20 (1959)
14. Church, A.: Review of A summary of symbolic logic and its practical applications by Edmund C. Berkeley, Review of A summary of symbolic logic and its practical applications by Edmund C. Berkeley, second printing. The Journal of Symbolic Logic 18(1), 68 (1953)
15. Church, A.: Additions and correction to a bibliography of symbolic logic - Index of subjects. The Journal of Symbolic Logic 3(4), 193–204 (1938)
16. Church, A.: Additions and correction to a bibliography of symbolic logic - Index of reviews. The Journal of Symbolic Logic 3(4), 204–206 (1938)
17. Church, A.: Additions and correction to a bibliography of symbolic logic - Index of authors. The Journal of Symbolic Logic 3(4), 206–212 (1938)
18. Church, A.: Review of Theoria algebrica dei meccanismi automatici by Grigore Constantin Moisil. The Journal of Symbolic Logic 36(3), 547 (1971)
19. Church, A.: Review of A relay machine for the demonstration of symbolic logic by W. Mays, D.G. Prinz. The Journal of Symbolic Logic 15(2), 138 (1950)
20. Comey, D.D.: Review of Philosophical problems of many-valued logic by A.A. Zinoviev. The Journal of Symbolic Logic 28(3), 2552–2560 (1963)
21. Feigenbaum, E.A.: Soviet cybernetics and computer science - 1960. IRE Trans. Electronic Computers EC-10, 759–776 (1961)
22. Further Citations, "Reviews of work on logic". The Journal of Symbolic Logic 32(4), 550–555 (December 1967)

23. Goodman, S.: The origins of digital computing in Europe. Communications of ACM 46(9), 21–25 (2003)
24. Harland, A.: Review of Mathematical Logic and Automation by V.I. Shestakov. The Journal of Symbolic Logic 29(1), 51 (1964)
25. Harland, A.: Review of Punched card synthesis of sequential systems of multi-position relays by I.V. Shestakov. The Journal of Symbolic Logic 31(1), 140 (1966)
26. Hohn, F.: Some mathematical aspects of switching. The American Mathematical Monthly 62(2), 75–90 (1955)
27. Huntington, E.V.: Complete sets of postulates for the theories of positive integral and positive rational numbers. Trans. Amer. Math. Soc. 3(2), 280–284 (1902)
28. Huntington, E.V.: A second correction. Trans. Amer. Math. Soc. 35(4), 971 (1933)
29. Kalin, T.A.: Formal logic and switching circuits. In: Proceedings of the 1952 ACM national meeting (Pittsburgh), Pittsburgh, Pennsylvania, May 2, pp. 251–257 (1952)
30. Miller, D.G.: Postulates for Boolean algebra. The American Mathematical Monthly 59(2), 93–96 (1952)
31. Monro, C.J.: Review of the book Symbolic Logic by John Venn. Mind 6(24), 547–581 (1881)
32. Moore, E.F.: Review of Punched cards synthesis method of switching functions by V.I. Shestakov. The Journal of Symbolic Logic 24(3), 259 (1959)
33. Mostowski, A.: Review of Representation of characteristic functions by expressions realizable by relay-contact circuits by V.I. Shestakov. The Journal of Symbolic Logic 12(4), 135 (1947)
34. Nelson, E.J.: Note on a recent set of postulates for the calculus of propositions by E.V. Huntington. The Journal of Symbolic Logic 4(2), 90 (1939)
35. Nelson, E.J.: Review of Application of Boolean algebra to switching circuit design and error detection, by D.E. Muller. The Journal of Symbolic Logic 20(2), 195 (1955)
36. Patterson, G.W.: Review of The relations between symbolic logic and large-scale calculating machines by Edmund C. Berkeley. The Journal of Symbolic Logic 17(1), 78 (1952)
37. Patterson, G.W.: Review of the paper by Robert E. Staehler An application of Boolean algebra to switching circuit design. The Bell System Technical Journal 31, 280–305 (1952); The Journal of Symbolic Logic 18(4), 348 (December 1953)
38. Patterson, G.W.: Review of Logical Machines - New light on W. Stanley Jevons by W. Mays, D.P. Henry. The Journal of Symbolic Logic 18(1), 69 (1953)
39. Pawlak, Z.: Review of An algebra of two-terminal circuits constructed exclusively of two-terminal components (Algebra of A-circuits) by V.I. Shestakov. The Journal of Symbolic Logic 21(4), 399 (1956)
40. Pawlak, Z.: Review of The modeling of the propositional calculus by means of simple four-terminal networks by V.I. Shestakov. The Journal of Symbolic Logic 22(3), 332–333 (1957)
41. Peckhaus, V.: Hugh MacColl and the German algebra of logic. Nordic Journal of Philosophical Logic 3(1), 17–34 (1999)
42. Rose, A.: Review of A dual arithmetic Interpretation of the 3-valued propositional calculus utilized in the simulation of this calculus by relay-contact networks, by V.I. Shestakov. The Journal of Symbolic Logic 37(1), 187 (1972)
43. Shannon, C.E.: A Symbolic Analysis of Relay Networks, Massachusetts Institute of Technology, MSc Thesis (1938)
44. Shannon, C.E.: The synthesis of two-terminal switching circuits. Bell Syst. Tech. J. 28(1), 59–98 (1949)

45. Shannon, C.E., Moore, E.F.: Machine aid for switching circuit design. In: Proc. IRE, vol. 41, pp. 1348–1351 (1953)
46. Stabler, E.R.: Boolean representation theory. The American Mathematical Monthly 51(3), 129–132 (1944)
47. Stelzner, W.: Review of Ocerki sotsialnoj istorii logiki v Rossii (Sketches of the social history of logic in Russia) by V.A. Bazhanov. The Bulletin of Symbolic Logic 10(3), 421–423 (2004)
48. Turquette, A.R.: Generalizable Kleene logics. Proc. American Mathematical Society 20(2), 361–367 (1969)
49. Wakulicz, A.: Review of On the transformation of a periodic sequence into a recursive one by V.I. Shestakov. The Journal of Symbolic Logic 22(2), 220 (1957)
50. Wakulicz, A.: Review of An algebraic method for the synthesis of switching circuits by V.I. Shestakov. The Journal of Symbolic Logic 22(2), 221 (1957)
51. Wakulicz, A.: Review of An algebraic method for the synthesis of autonomic systems of two-positional relay by V.I. Shestakov. The Journal of Symbolic Logic 23(2), 231–232 (1958)
52. Ware, W.H.: Soviet computer technology - 1959. Communications of ACM 3, 131–166 (1960) also in IRE Trans. Electronics Computers EC-9, 72–120 (March 1960)
53. Watanabe, H.: Activities on circuit theory in Japan. IEEE Trans. Circuits and Systems CAS-31(1), 112–123 (1984)
54. Wiener, N.: Certain formal invariances in Boolean algebras. Trans. Amer. Math. Soc. 18(1), 65–72 (1917)

Acknowledgment

Authors are grateful to many friends and scholars that helped in providing the litearture and giving advices and comments.

We gratefuly acknowledge the help and advices by Professors **Claudio Moraga**, **Franz Pichler**, **Tsutomu Sasao**, **Akihiko Yamada**, and **Heinz Zemanek**.

Special thanks are due to
Mrs. **Marju Taavetti** and Mrs. **Pirkko Ruotsalainen**
of Tampere Univeristy of Technology, Tampere, Finland
for the help in collecting literature.

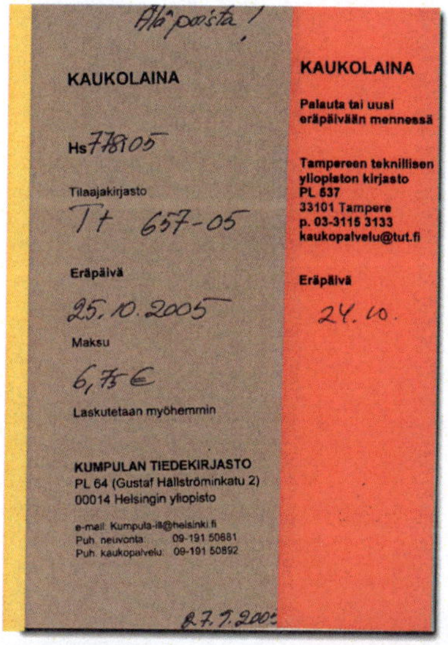

KAUKOLAINA

Hs

Tilaajakirjasto

Tt 657-05

Eräpäivä

25.10.2005

Maksu

6,75 €

Laskutetaan myöhemmin

KUMPULAN TIEDEKIRJASTO
PL 64 (Gustaf Hällströminkatu 2)
00014 Helsingin yliopisto

e-mail: Kumpula-ill@helsinki.fi
Puh. neuvonta 09-191 50681
Puh. kaukopalvelu: 09-191 50892

KAUKOLAINA

Palauta tai uusi eräpäivään mennessä

Tampereen teknillisen yliopiston kirjasto
PL 537
33101 Tampere
p. 03-3115 3133
kaukopalvelu@tut.fi

Eräpäivä

24.10.